王劲韬 著
By Wang Jintao

Scheme Drawing

方案制图

景观设计手绘教程
Hand Drawing for Landscape Design

中国建筑工业出版社
CHINA ARCHITECTURE & BUILDING PRESS

前　言

　　本册以工程方案制图所需的各类图纸为中心，针对不同尺度下的图纸平面表达、局部剖面表现、立面多层次透视表达、总体鸟瞰以及以上三类图纸综合运用的设计表达技巧。

　　在平面图部分，重点示范了多层次植物平面构图、色彩原则与技巧，针对色彩冷暖、阴影设置、树形对比以及马克笔与钢笔、彩铅等工具的综合使用等方面进行示范；对于大尺度规划性质的景观平面强调用多种图式、图标以及文字辅助表达的重要性，提出清晰准确、美观兼顾的要求，对于大尺度景观所涉及的多层次信息叠加易产生混乱等问题，提出采用不同覆盖率的多层次材料表现，以便内容层层相扣，效果保持整洁清晰。

　　剖面图作为设计过程中重要的工作草图比平面表达具有更大的选择优势，本册以自然风景及滨水驳岸、广场等项目为例，针对项目的不同特色提出相应的表现策略。

　　大众尺度的鸟瞰表现是最近几年国内相关专业学生考试快题与设计专业课中的薄弱环节，针对鸟瞰透视与传统轴侧图纸在表达方面的优势对比，本书提供了包括平面拉伸、Google 图叠加与图片改绘等多种形式的鸟瞰图制作技法与示范。最终目的还是整体协调，在充分表达设计意图的前提下，突显手绘图纸的艺术性。

Foreword

Focusing on various kinds of drawings needed by engineering proposals in the second volume of this series, this volume illustrates drawing plane expression, local section expression, multilevel elevation perspective expression, overall aerial view under different scales and the design expression skill of the comprehensive application of the above three kinds of drawings. The plane legend section mainly demonstrates the plane composition, coloring principle and skill of multilevel plants, with demonstration in aspects of cold/warm colors, shadows, comparison of tree forms and the comprehensive utilization of tools, like marker, fountain pen and color pencil; for landscape plane of macro-scale planning nature, the importance of various schemas, icons and auxiliary verbal expressions is emphasized along with the principle of giving considerations to both precision and aesthetics. For the confusion caused by the overlying of multilevel information involved in macro-scale landscape planning, the method of using multilevel materials with different coverage scale with closely connected contents to keep the design tidy and clear is proposed.

As an important base map in the design process, Section expression has a greater selective advantage than plane expression. Taking the projects of natural scenery, waterfront revetment and square as examples in this volume, relevant expression strategies are proposed as per the different characteristics of projects.

In recent years, the aerial view expression of public scale has been a weak link in the domestic examinations for students of relevant majors and in the specialized courses for design major. Based on a comparison of expression advantages between aerial view perspective and traditional isometric drawings, various types of techniques for making aerial view, including planar elongation, overlying and altered rendering of Google maps , are provided in this volume. The ultimate purpose is to feature the artistic qualities of hand drawings under the premise of fully disclosing design intent with an overall coordination.

目 录

008	第 1 章	**平面图的表达**	
008		1.1 中等尺度比例的平面表达	
022		1.2 大尺度比例的平面表达	
052		1.3 景观规划平面概念表达	
064	第 2 章	**剖、立面图的表达**	
064		2.1 局部尺度的剖、立面图	
076		2.2 整体尺度的剖、立面图	
084	第 3 章	**鸟瞰图的表达**	
086		3.1 鸟瞰图步骤	
104		3.2 鸟瞰图特殊表达技巧	
114	第 4 章	**景观多义表达**	

Contents

008   **Chapter I**   **Expression of Plans**
008   1.1 Expression of plans in medium scale
022   1.2 Expression of plans in large scale
052   1.3 Concept expression of landscape planning plan

064   **Chapter II**   **Expression of Sections and Elevations**
064   2.1 Section and elevation of local scale
076   2.2 Section and elevation of integral scale

084   **Chapter III**   **Expression of Aerial View**
086   3.1 Steps of aerial view
104   3.2 Special expression techniques of aerial view

114   **Chapter IV**   **Polysemous Expression of Landscape**

第1章
平面图的表达
Chapter I
Expression of Plans

1.1 中等尺度比例的平面表达
1.1 Expression of plans in medium scale

1:500 以上平面图表现

乔木和灌木可以多层叠压，相互掩映，用颜色及冷暖对比来表达其空间关系，使之结构清晰，层次丰富。"三五成群"的群树多冷暖平衡。乔木用色偏冷，灌木则偏暖，如蓝紫配土黄或红褐；乔木用色偏暖，灌木用色则偏冷，如棕或暖灰配墨绿等，以达到视觉补色上的一种色彩平衡。

在草地的表达上可用钢笔排线来表现地形，也可画些草点来表现质感肌理。

Expression of Plan above 1:500

Arbors and shrubs can adopt multi-level overlying, which sets off one another. The comparison of cold and warm colors can be used to express their spatial relationship so as to clarify the structure and enrich the gradation. Clustered trees shall be balanced in color. If arbor adopt a relatively cold color, shrubs shall adopt warmer colors. For example, bluish violet can be matched with yellowish brown or reddish brown; if arbors adopt a relatively warm color, shrubs shall adopt colder colors. For example, brown or warm grey can be matched with blackish green to achieve a color balance in visual sense with complementary colors.

As for the expression of grassland, landform can be expressed by rows of lines drawn with pen, and grass can be drawn to promote the texture of the land.

锥形的投影反映出植物实体（笔柏）的形状
Tapered projection reveals the shape of the plant (Sabina chinensis).

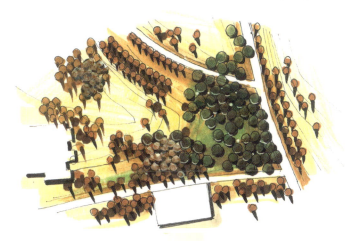

以下 5 例表现干燥坡地的栽植景观配置。大量用暖色粗线条彩铅，用浅棕、灰画出乔木，用更深的色彩点出灌木、景石，用最深的黑标出投影，上下叠压的层次感突出，使平面图也能表现出环境特色和场地感。

手绘的随意、模糊性表达多种意向的可能性在此类平面中充分展现，往往由此类平面图直接引出立面甚至小幅透视效果，对设计意图的综合展示极有帮助。

The configuration of transplant landscape on dry sloping fields is shown with the below five examples. Warm color pencils with thick lines are used heavily: light brown and grey are used to draw arbors; darker colors are used to dot shrubs and scene stones; darkest black is used to mark out the projection. The technique of vertical overlying highlights a sense of gradation, making it possible to reveal the environment characteristics and sense of field with plan.

The possibility of multiple intentions bought by the arbitrary and obscure expression of hand drawing is fully displayed in this type of plan. Elevation and even small-scale perspective effects tend to be drawn forth directly from this type of plan, which is extremely helpful for the comprehensive demonstration of the design intent.

特殊的植物平面表达
Special expression of plants plan

"勾"出两个层次
Draw two levels

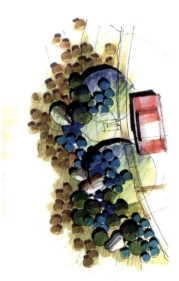

上下重叠，同时表达两个层次的细节
Render the details of two overlapped levels

涂改液"涂"出两个层次
"White out" two levels with collection fluid

当图的尺度很小的时候,主景树可以简略到只画一个圆来更加细致地刻画林下空间。树的明度不宜过强,用色彩的浸透变化以及阴影去表现形体和空间,有些地方可以用高光笔或涂改液点染一边,使结构层次表达得更清晰。

When the scale of the drawing is small, the feature tree can be simplified to a circle and the space under the tree shall be drawn with meticulosity. The lightness of the tree shall not be too bright, and the shape and space shall be rendered through the change of colorific permeation and shadow. Some places can be polished with highlight pen or correction fluid to make the expression of structural layer clearer.

（1）树冠之间要相互避让。

（2）树圈之间可以相互叠压，要注意表现层次。大型阔叶树可以只画树的轮廓阴影，着重表现树下空间。

（3）用颜色色相的冷暖和明度把空间层次拉开。

(1) There shall be no overlying of canopies.

(2) Tree circles can overlap each other as long as attention is paid to the manifestation of levels. Large broad-leaved trees can be simplified to the outlines and shadows of the trees, with emphasis on the space under the trees.

(3) The space levels shall be highlighted by using colors with different tones and lightness.

植物平面步骤图

画一片沙滩,金与黑,如何突出金色沙滩的质感?从海边防护林到水面的冷暖变化对比使得"沙滩更金黄""海面更碧蓝"。沙滩越靠外颜色越浅,由于黑色森林和蓝色水面"切割"作用,临近水边的沙滩几乎用彩铅轻涂一层即能显示出"金色"。

Process of drawing plans

When drawing a beach of golden and black, how to make the texture of the golden beach more prominent? The contrast produced by the change of color tones from the seaside protection forest to the water surface makes the golden beach and the blue sea look more vivid. Due to the incision between the black forest and the blue water, the beach near the water can be rendered golden with a light coating of color pencil.

此类草图要注意特色植物的表达，点缀特色景观元素，如小船、遮阳伞等，用排线和投影加强防护林的层次，是极为迅速有效的表达方法，通过着色更进一步说明场地性质。

For this type of hand drawing, attention shall be paid to the expression of featured plants and the embellishment of featured landscape elements, like boat and beach umbrella. An extremely rapid and effective expression method is to increase the levels of the protection forest with rows of lines and projection, and to further illustrate the nature of the site via coloring.

（1）完善墨线图，在CAD的线图基础上，加画树的投影等细节，分出画面前后层次。

（2）按照一定的顺序上颜色，由上到下或由左到右，此类平面作图一般不用先浅后深，大面积浅底色用彩铅"刷"出，深色树丛用马克笔点出。

(1) Complete the ink drawing; on the basis of CAD diagram, details (such as, the projection of the trees) shall be added to make the front and back levels of the drawing more prominent.

(2) The plan shall be colored as per a certain consequence, be it from top to bottom or from left to right. This type of plan normally doesn't adopt the method of rendering colors from light to dark. Large areas of light bottom color shall be brushed with color pencil, and dark bush shall be dotted with marker.

下图为典型的小庭院表达，平面线稿底图较为复杂，仅景石一项，就有无规则的石、半规则的石、汀步和完全自然化的石滩石等，植物则包括乔、灌、地被层层覆压，几乎密不透风。所有的秩序，条理均是由后期的色彩所规范，最强烈的对比由日本庭院两大特色——白沙和苔藓地衣组成，次级对比为槭树，木头板等中间层次，大量的日本松用深色被衬托，层层叠加，多而不乱。

The drawing below is a typical expression of a small garden. The base drawing of the plan sketch is relatively complicated. Take scene stone for example, there are irregular stones, semi-regular stones, stepping stones and totally naturalized rock bench stones. As for plants, arbors, shrubs and ground covers overlap each other layer upon layer, making it almost impenetrable. All orders and systems are regulated by colors added in later stage. The strongest contrast is made by the two biggest characteristics of Japanese garden —— the white sand and the moss; the secondary contrast is the middle level of maples and wood plates; set off by dark background, abundant and overlying Japanese pines look rich but not crowded.

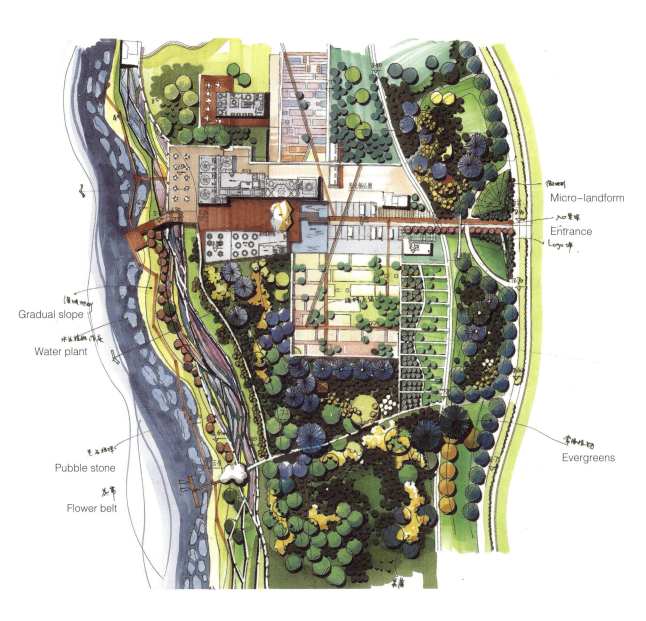

　　上图用鲜艳明快的色彩表现大面积的花田、花带。大面积颜色要统一避免杂乱无章，小面积可适当应用补色，以达到相互映衬色彩平衡的效果。如左图花带为紫色系，运用浅紫、粉红、玫瑰为主色调，用黄色（紫色的补色）来点缀其间，统一而富于变化。

　　Bright and lively colors are used to render large areas of flower fields and flower belts on the above drawing. Colors of large areas shall be unified to avoid disorder and messiness, and complementary colors can be properly used in small areas to achieve a balanced and mutually complementary color effect. For example, the flower belt in the left-hand drawing is of purple color scheme. Light purple, pink and rosy are used as the main tones with yellow (complementary color of purple) embellished among them, achieving a unified yet changeful effect.

在大尺度的平面图上，更要把握住整体的感受，树圈可以画得很简省，成片的群树色彩要统一，避免每棵树的颜色变化过多而导致的杂乱无章。投影的形状是很容易被忽视的地方，借此我们却可以来对不同树种加以区别，如阔叶树弯弯的"月牙形"和针叶树俏丽的"锥形"投影使平面图和鸟瞰图更加直观。

For plan of large scale, it is more important to grasp the integral feeling. Tree circle can be simple, and the color of clustered trees shall be unified to avoid the disorder and messiness caused by too many changes in color. The shape of projection, whereby can be used to distinguish different varieties of tress, is a point that is prone to be neglected. For example, the arched "crescent-shaped" projection of broad-leaved tree and the pretty "cone-shaped" projection of needle-leaved tree help to make the plan and aerial view more intuitive.

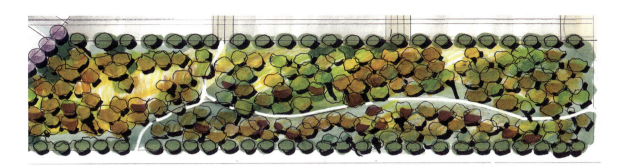

1:1000以上的大型场地平面多着重表现设计构成，场地特征和植物群落特征，往往不再对一树一圈作过细描绘。

在尺度稍大的平面图中，树圈的数量比较大，适宜成片描绘。群树和行道树的组合关系可以平涂、过度相互晕染，表现光影颜色的变化而不用将每一棵树细致刻画，投影也整体点染，草坪部分可大量留白增强画面的虚实对比，正是"突出重点，疏密得当"。

由此，整体层次关系从前述的一组植物扩展到一片场地，前述树圈之间的对比关系同样适用于整体树林、草坪之间的大范围对比——虚实得当，对比明确即可。

The plan of large-scale site with scale above 1:1000 usually focuses on the expression of design constitution, site features and features of plant community, and detailed depiction of each tree and each circle is left out in most cases.

For plan with relatively large scale and relatively large quantity of tree circles, it is suitable to paint in areas. The combination relation between clustered trees and border trees can be rendered via flat painting and excessive mutual blooming to show the change of shadow and colors without too detailed depiction. Projection can be polished as a whole. A lot of blank spaces can be left for the lawn to enhance the virtual-real comparison of the drawing, which as a saying goes, makes the focal points stand out with proper density.

Therefore, the overall hierarchical relation expands from the above-mentioned a group of plant to a patch of land. The above-mentioned contrast between tree circles also applies to the contrast of larger areas, such as the woods as a whole and the lawns, as long as it has a proper and explicit virtual-real comparison.

某大型农场、酒庄的整体规划

　　这张图是在风景园林新青年的视频录制时所做,全部用时 45 分钟左右,作图过程至今在网络上能够找到,做这样一个实验,目的是告诉全国的学生和设计师,一个小时之内可以完成的图面工作量。事实上,我们在工作中这个创作过程完全可以大大延长,但是,即便最优美的平面图也可以在短短的一、两个小时之内完成,而绝对不需要用一两周时间做所谓的多层次渲染。

Overall planning of a large farm and chateau

　　This drawing is finished in 45mins and the drawing process is recorded by Youth Landscape Architecture, which can still be founded on the Internet. The purpose of such an experiment is to show the drawing workload that can be completed in an hour to the students and designers. In fact, the production process can be greatly extended in our actual working process. But, even the optimal plan can be finished in one or two hours. It definitely does not require one to two weeks for the so-called multi-level rendering.

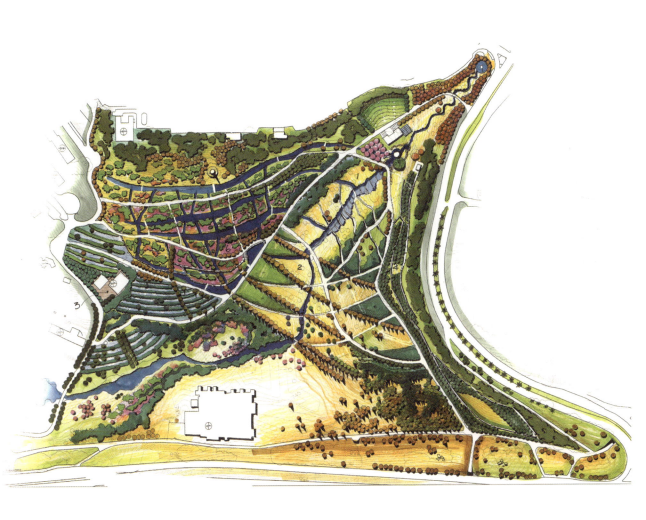

　　本图的黑、白、灰三个层次，没有使用覆盖，而是用空的方式，完全空白的建筑、道路，半透明的土黄色彩铅做出了灰色，中等覆盖力的油性马克，在整体土黄色彩铅基础上，覆盖出了画面的主体——丛林和种植园。画面没有一个多余的层次，同样也不需要所谓的分层、渲染、退晕。总之将快和好结合到最佳程度，不是时间的问题，而是多种技巧的最佳搭配，和对材料性能的充分利用。

　　There are three levels — black, white and grey — in this drawing. It does not adopt the method of overlying but the method of leaving blanks. The buildings and roads are totally blank. Grey is rendered via semi-transparent yellowish brown color pencil. Oily marker with medium covering power is used to cover the main part of the drawing, namely the jungle and the plantation, on the basis of the overall yellowish brown color pencil. There is no redundant level in the drawing, and it does not require the so-called layering, rendering and color retreating. In a word, it's not a matter of time to combine speed and quality to the optimum degree, but a matter of optimum match of various skills and utmost utilization of material property.

1.2 大尺度比例的平面表达
1.2 Expression of plans in large scale

　　这类大尺度、小比例的草图，目的只有一个：用多种技术手段，各种图式、图标甚至文字帮助阐释，展现设计概念与思考。这类图纸同样需强调美感，而且以图面的清晰可读性为第一要求。上图为唐山植物园、苗圃和中央湖区的布局总图，涉及三类景观基质，苗圃、景区和外围的绿色维护，所有的道路用色笔覆盖而成，景观构筑以及植物用马克笔绘制，空出的部分即为水面，色彩从冷到暖，层次分明。但这一切均不是为了美观，注意"清晰表达"在此阶段永远是第一位考虑的因素。

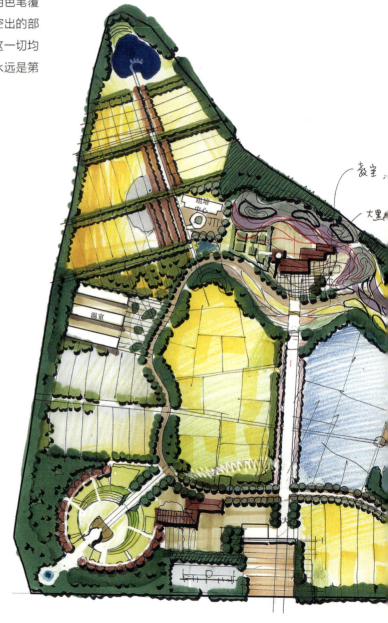

For this type of sketch with large scale and small proportion, the only purpose is to reveal the design concept and reflection with multiple technological means and with the help of various schema, icon, and even words. This type of drawing also emphasizes aesthetic perception and clarity, with clarity and readability of the drawing as the primary requirement. The above drawing is a general plan of the Tangshan Botanic Garden, the nursery garden and Central Lake District, which involves three types of landscape matrix. Color pencil is used to cover the green maintenance of the nursery garden, the scenic region and the periphery as well as all roads. Marker is used to drawn the landscape structures and plants. The blank part refers to the water. The color changes from cold tone to warm tone, revealing a distinct gradation. However, all are not for the sake of aesthetics. Please remember that "clarity of expression" shall always be the first concern in this stage.

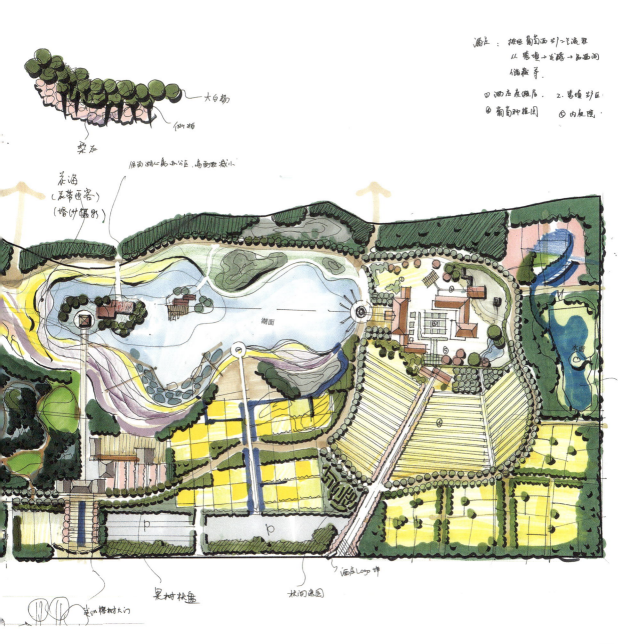

本页和下页两张图反映所谓的一草与二草的区别，一草用黄色草图纸直接涂抹，这里面有个技巧：黄色草图纸可以分辨出油性马克和水性马克以及普通记号笔的色彩层次，同样色度的工具，油性马克最浅，水性马克次之，普通记号笔可以在草图纸上画出和普通白纸上一样深度的线条。所以，本图用油性马克画出了大面积的浅色疏林草坪，在其上覆盖叠加水性马克的蓝色湖面，最后在一张几乎满满当当的图面上用记号笔画出了水线和道路。右侧黄色草图其实较为潦草，但作为第一阶段的草图深度已经足够。左侧的二草实际是在基础上将外轮廓墨线清晰化，将一草中的模糊区域赋予特定功能，以及增加部分细节，仅此而已。两层图纸不仅前后相继，而且在必要的时候可以进一步叠加，形成更为复杂的图面效果。如前页企业园平面所表现的那样。

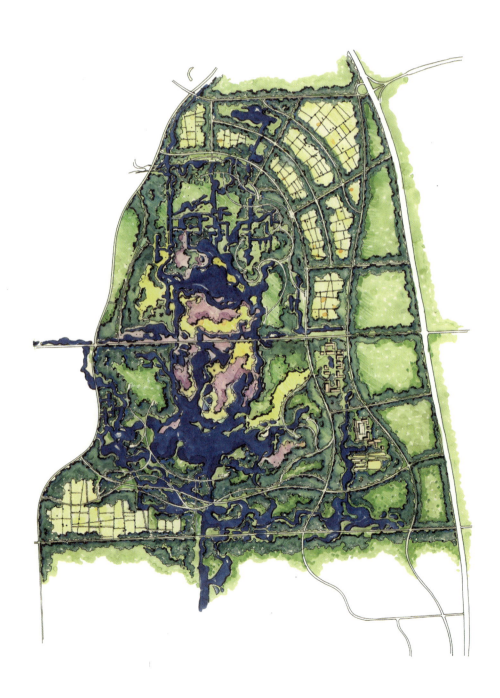

  The two drawings reflect the so-called distinction between the first draft and the second draft. For the first draft, the designer directly draws on a piece of yellow sketching paper. There is a skill here: yellow sketching paper can distinguish the gradation of oil-based maker, water-based marker and common marker. For tools with the same color scale, the color of oil-based marker is the lightest, followed by water-based marker, and common marker can draw lines of the same darkness on the sketching paper as on common white paper. Therefore, in this drawing, oil-based marker is used to draw large areas of light-colored sparse woodland lawn, on the top of which water-based marker is used to draw blue lake surface. In the end, common marker is used to draw the waterline and roads on the almost fully filled drawing. The yellow sketch on the right side is actually relatively rough-and-ready, but it is detailed enough for a sketch in the first stage. The second draft on the left side is actually made on the basis of the first draft. The lines of the outer contour are further clarified, and specific functions have been given to the fuzzy areas on the first draft. In addition, partial details are added. That's all. These two drawing are not only made one after another. If necessary, they can be further overlaid to form more complicated effect of drawing, as shown on the Industrial Park Plan on the previous page.

草图阶段的图纸表现可以非常多样，以基本满足清晰美观和尽可能多地容纳规划设计信息为目标，本图即为多层叠加以后自动形成的一张景观总图，虽然信息繁多，但做到层层相扣，并不显得杂乱。

多层信息叠加的分析图纸通常可以分成道路交通（车、人、慢行等系统）、竖向信息（包括土方排水等信息）、功能区块等，这些步骤可用草图纸完成，要能够叠加，反映不同层次的规划诉求。

The expression of drawing in the sketching stage can be highly diversified. With the aim of basically satisfying the requirements on clarity and aesthetics and containing as much design and planning information as possible, this drawing is a landscape general drawing automatically formed with the overlying of multiple layers. Although it contains numerous pieces of information, it does not look messy because of the clear hierarchy.

Analysis drawing with the overlying of multilayer information can normally be divided into road traffic (system of vehicle, people, lose headway, etc.), vertical (including information of earthwork and drainage), functional block, etc. These steps can be done on sketching papers. They shall be capable of overlying to reflect the planning demands of different layers.

鄂尔多斯伊金霍洛旗大南沟水库周边景观规划
Landscape planning of the surrounding of Danangou Reservoir in Yijinhuoluo Banner, Erdos City

本页为最简易的黄色草图纸所做第一轮分析，涵盖内容较多，包括了主次干道划分、水系布局、景观绿廊水体和各个主要观景点设置等。

草图需要明确区分各区域的相互关系，包括图底关系、不同功能区分。比如水体用色最深，绿带次之，广场、草坪、和农田（城市农庄）等用地颜色最浅；所有景观主次道路系统全部用高覆盖力的涂改液在黄色纸上一次绘出，清晰准确。

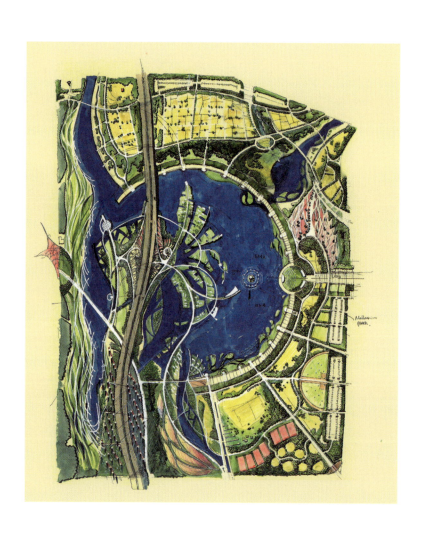

This page shows the first round analysis of the simplest yellow sketching paper. It contains numerous contents, including the division of primary and secondary trunk road, the arrangement of water system, landscape and pergola water and the setting of each major scenic lookout.

The sketch shall explicitly distinguish the mutual relation between each area, including the figure-ground relation and the distinction of different functions. For example, the color of the water shall be the darkest, followed by the green belt, while the color of the square, lawn and farmland (urban homesteading) shall be the lightest; all primary and secondary road system of the landscape shall be clearly and accurately drawn on yellow paper with correction pen of high covering power once and for all.

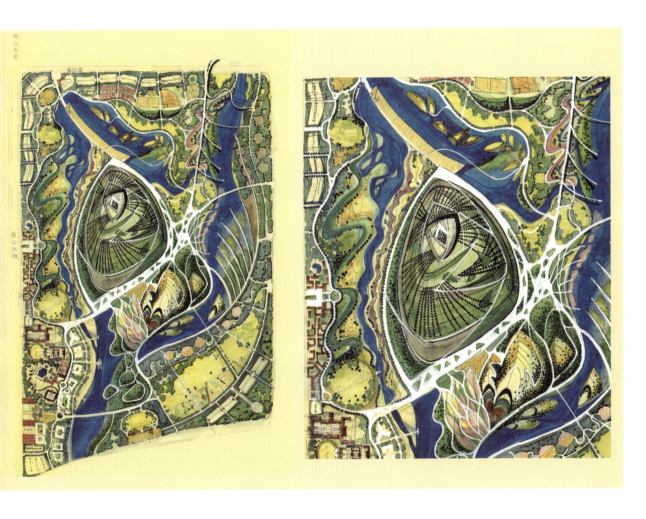

唐山南湖植物园规划

多次修改，边设计，边表达。极大发挥了草图设计的灵活性、多义性，并配有大量文字说明及注释，将设计所思、所想尽可能全面地表达在同一个平面之中。

以下两例，为唐山南湖植物园及世界园艺展场地规划，左图为园艺展中心场地，右图为整个植物园区域规划，两者相互对照，逐层深入，草图采用线、色块和涂改液多层叠压。在做大尺度的规划方案时，白色的路网和蓝色的水系相互穿插，成片的草地绿带被分割成优美的形状，留白的建筑点缀其间，灵活运用色彩表现场地性质，丰富而有条理，形成了具有律动的音乐美感。

Planning of Southern Lake Botanical Garden in Tangshan

While drawing the sketch, the design can be modified for several times. The expression is shown along with the design. Sketch design is highly flexible and polysemous. With plenty of explanatory notes and annotations, the design intent and philosophy can be shown as comprehensive as possible on the same plan.

The below two drawings are the plans respectively for a botanic garden and for the site of an international horticultural exposition. The drawing on the left is for the central site of the international horticultural exposition, and the drawing on the right is the overall planning for the botanic garden area. These two drawings are cross referenced with overlying layers. The sketch adopts multilayer overlying of lines, color blocks and correction pens. While making planning scheme with large scale, the white road network intersects with the blue water system. Stretches of grassland and green belt are segmented into exquisite shapes, dotted with blank buildings. Colors are used flexibly to demonstrate the nature of the site. It is rich and clear, forming a rhythmic and aesthetic feeling of music.

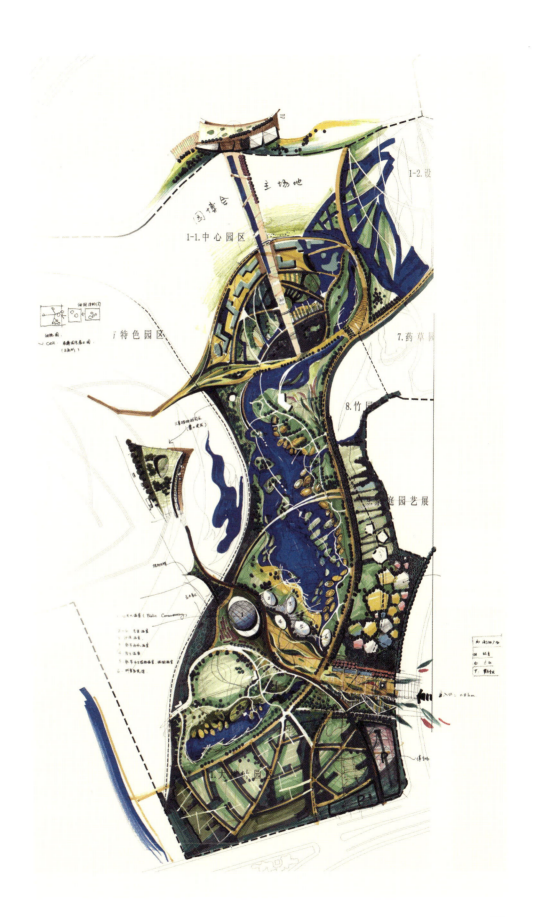

本页设计为唐山世园会最初选址的第一轮方案，场地基质条件极差，由一座煤矸石山、一片污染较为严重的水域以及开滦煤矿所属的大面积重污染的工业棕地组成。规划的目标是要造就一座美丽的山体，净化一座清洁的湖面，然后在大面积的工业棕地上利用生态手段营造出数十座企业园园区。在一草阶段对于园区的形态，引入了树叶园、细胞园、流线形的间歇性湿地的各种形态。就图面表达而言，用了高低两种覆盖力的马克笔，以及墨线将画面尽可能丰富化，道路完全用涂改液覆盖而成，整张 A0 的图纸包含了多层次的形式和大量的图元信息，全部工作量在六小时之内完成，实现了高度快速化、清晰化和系统化的工作逻辑。与同类的电脑图纸相比，其优势在于可以随时就创作灵感出现，形成多选择的方案修改。如图右上角针对大树叶园的形式同时提供的两套方案，对于它的叠加表现，只需要用墨线引出即可，异常简单。当然，更常见的方式是用固体胶贴上半透明的草图纸，可以随时揭开第一方案盖上形成第二方案。其叠加原理与电脑的 ps 图层方式完全一样，方便程度丝毫不减。

The design on this page is the first round scheme of the primary site of Tangshan International Horticultural Exposition. The basic condition of the site is extremely unfavorable. It consists of a gangue mountain, a stretch of badly polluted water and a large patch of heavily polluted industry brown field belong to Kailuan Coal Mine. The purpose of the planning is to beautify the mountain, purify the lake and build tens of corporate parks on the large patch of industry brown field via ecological means. In the stage of the first draft, various forms of intermittent wetland, including leaf garden, cell garden and streamline, are introduced as garden forms. As for the drawing expression, two kinds of markers with different covering power and inked lines are used to enrich the drawing as much as possible. The roads are entirely covered with correction pen. The whole piece of A0 size drawing contains multi layered forms and numerous graphic element information. The whole workload is completed in six hours, realizing a logic and systematic control of work with high velocity and clearness. Comparing to computer drawings of the same kind, its advantage lies in that multiple modification plans can be formed in no time as new inspirations pop out. As shown in the drawing at the top right corner, two schemes are provided for the large-leaved garden. To show their overlying expression, one only need to drawn forth the inked lines, which is extremely easy. Of course, the commoner way is to stick a semitransparent sketching paper with solid gum on the top of the drawing. The first scheme is revealed if the paper is uncovered, and the second scheme is formed if the paper is covered on top. The superposition principle is exactly the same as that of the PS coverage on the computer, and it is also very convenient.

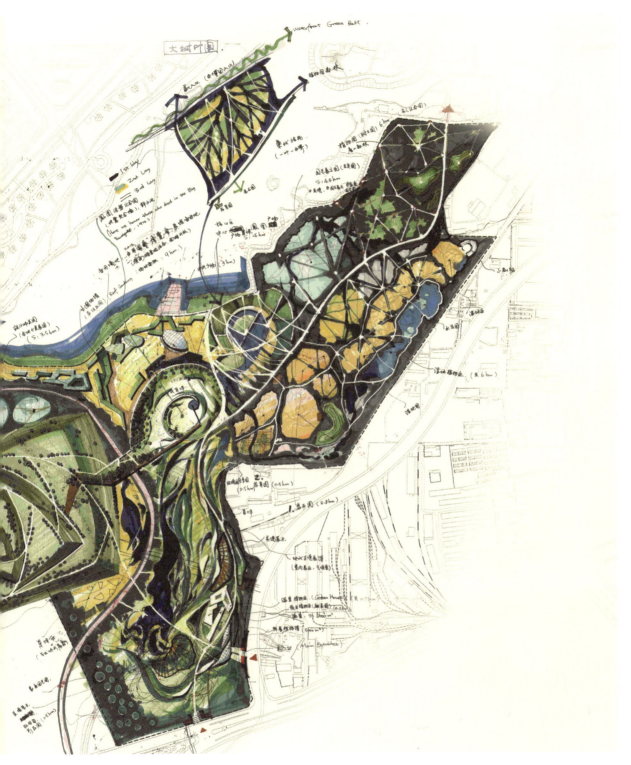

唐山南湖植物园总体规划

本页及后页两张图是作者在十余年前学生时期的设计作品，整体效果清晰、浑厚，在总的灰调子基础上加入了丰富的层次，可看性较强，可读性也较强。图面扭转了90°角，故图中所有建筑均为正南北向。

The drawings of this page and next page is designed by the author's students in about ten years ago. The overall effect is clear and vigorous. Rich gradation has been added to the overall grey basis, making it worth-seeing and worth-reading. The drawing has been rotated for 90°, therefore the buildings in the drawing all face due north and south direction.

山东青岛即墨某风景名胜区外围缓冲地带的村镇改造及度假区景观规划
Village and small town renewal and resort landscape planning of the buffer zone on the periphery of a scenic spot in Jimo, Qingdao, Shandong Province.

就图面表达而言，这是一张所谓最拘谨的平面图，在这里列出是希望它能够成为初学者的一个样板。在设计的能力欠缺把握不强的情况下，可以用工具的选择加以弥补。比如第一覆盖力的彩铅，不仅可以多次修改，而且各色彩铅的混搭叠加，能体现出近似模糊的表达效果。当这一切表达变得清晰肯定以后，你完全可以在此基础上覆盖一层肯定清晰的水性马克笔。如此一来，即便最没有把握的设计者，也能画出看上去肯定、毫无慌乱的作品。

In terms of drawing expression, this is a so-called the most formal plan. The reason why I put it here is that I hope it can be a sample for the beginners. When one does not have a strong mastery of design ability, proper tools can be selected to compensate for it. For example, if color pencil with low covering power is selected, it is possible to make multiple modifications. In addition, the mix and overlying of different colors made by color pencil can present a blurry expression effect. After all expressions become clear and definite, you can cover a layer of definite and clear water-based marker on that basis. As a result, even the least confident designer is able to draw a work that looks certain and composed.

本页两图选择了规划实践过程中,尺度差别最大的两个极端,左图为武汉磨山风景区全貌,尺度为1~2平方公里,右图为不足1公顷的庭园设计。在此做出对比,表现出手工的平面草图在不同尺度下的不同表现方式,左图的水面分为自然风景区的水面和生产性的鱼塘两部分,分别用深、浅两色表达。风景区核心地段为主要的设计区域,用暖色马克笔,外围缓冲区域用冷色的绿,体现出明显的内外两个区域。所有的景观构筑均弱化为空间中的一个点,其形式、尺度等问题都放到下一阶段解决,整个图面的创作目的,就是为下一阶段的深入设计以及分区规划确定一个确信无疑的整体框架即可。

在这一阶段必须集中所有注意力关注于结构和尺度,在此方面任何一点失误都会从整体上改变以后工作的方向。所以任何具体的建筑形式问题都不予考虑。

右图则正好相反,在景观元素表达主体极为有限并确定无疑的空间形式下,美学因素、色彩搭配、线条的曲直缓急等美学因素成为空间表现的主体。原则上讲,达到1:1000以下的所有设计草图都应该遵循这样的原则,比如右图的植物表达,分出了乔、灌、草的不同形式,大片的岩石园是直接用涂改液画出,建筑内庭院用色较为沉着浑厚,外部用色相对清淡单薄,显出层次上的差异性。

总之,两类尺度由于空间元素的差异,一个弱化了细节之间的变化,另一个则将空间的细节表现作为主体。目标都是在一张有限的二维平面图上,根据人的欣赏习惯,提供了最为适度的具有美感的可读性。这种元素过多,超越了人的辨识能力,则画面显得混乱,反之,过少则画面显得单调。我们所说的图面效果,其实就是提供最为适用的可读性,满足视觉的要求。

Here select two extremes with the largest scale distinction in the practice process of planning. The left drawing is a full view of Wuhan Moshan Scenic Zone with a scale of 1~2km^2, while the right drawing is a garden design less than 1hm^2. A comparison is made here to reveal the different expression methods under different scales of hand-drawn plan sketch. The water of the left drawing can be divided into two parts, the water of the natural scenic zone and the productive fishpond, which are expressed respective with a light color and a dark color. Markers of warm color is used to render the core area of the scenic zone, which is the major design region, while green of cold color is used to render the periphery buffer area, making the two inside and outside areas distinctive. All landscape structures have been weakened to a spot in the space, and the expression of their forms and scales all have been left to the next stage. The creation purpose of the whole drawing is to confirm an assured overall framework for the detailed design and zoning plan of the next stage.

In this stage, all attention shall be focused on the structure and scale. Any fault in these aspects can alter the direction of future work as a whole. Therefore, no consideration is given to any specific building types.

The right drawing is just the other way around. Under the condition of a highly limited landscape element expression subject and an assured spatial form, aesthetic factors, including aesthetic elements, color matching and routing of lines, have become the subject of spatial expression. In principle, all design sketches below 1:1000 shall follow such a principle. For example, for the plant expression on the right drawing, different forms of trees, bushes and grasses are distinguished, and blocks of rock garden are directly drawn with correction pen. Relatively composed and dark color is used to render the internal space of building courtyard, while relatively light and thin color is used to render the external space.

In short, due to the difference of spatial elements, for the drawings with two types of scales, one weakens the change of details; the other takes the expression of spatial details as the subject. The aim of both is to provide the most proper and aesthetic readability on a limited two-dimensional plan according to people's appreciation habits. If the quantity of elements surpasses people's identification ability, the drawing would look messy; otherwise, it would look monotonous. When we talk about the effect of drawing, we are referring to the provision of the most proper readability to satisfy visual requirements.

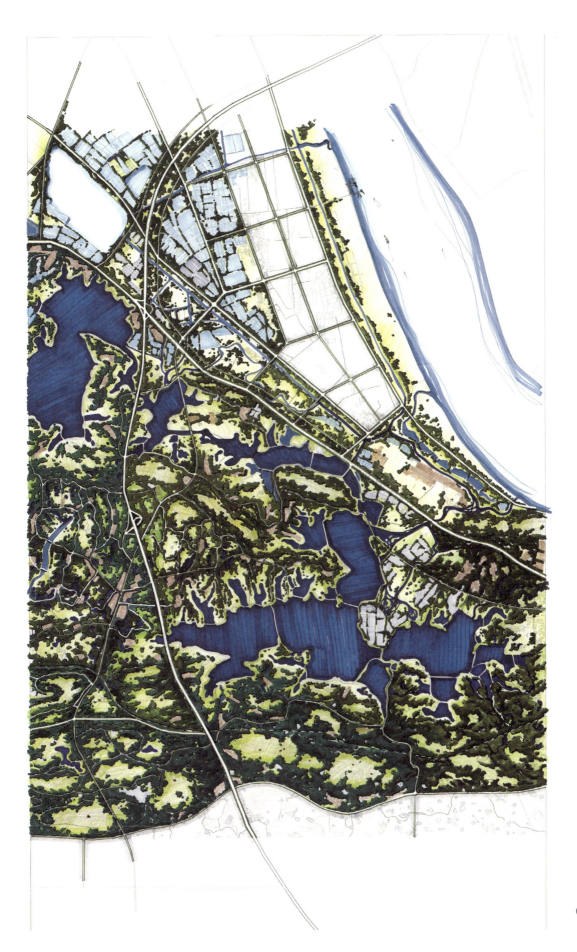

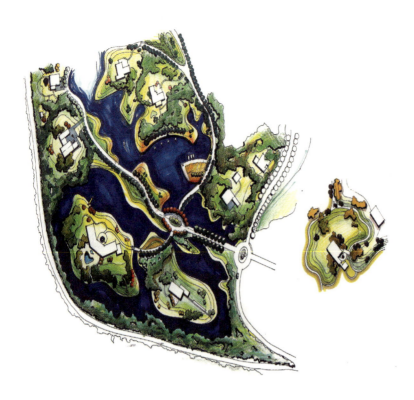

上海某别墅区的景观单元图面表现。图中的建筑全部留白，外围的植物全部用象征性的树丛处理，图面只关注了以下三点：

（1）空间的私密（公共会所与私有独栋别墅用色上冷暖对比）；

（2）空间的围合程度差别；

（3）水面与岛屿的形态对比。

The drawing expresses the landscape unit of a villa district in Shanghai. All buildings on the drawing are left blank, and all peripheral plants are expressed with symbolic grove. The drawing only pays attention to the three points below:

(1) The privacy of space (cold and warm colors are used to distinguish the public club and private villa);

(2) The difference of spatial enclosure degree;

(3) The comparison of water form and island form.

左图为武汉磨山风景区图纸的局部细化，表现了当图纸从 1:20000 到 1:2000 之间的图面表达的巨大差异。在总体规划图中的一根线变成了一条路，一片丛林甚至一个马克笔的笔触在此已经放大为一片农业园。此阶段所有的形式和空间要素都开始发挥作用，水面分出深浅，块田分出了冷暖，岛屿分出了各种形式的连接与道路。这一阶段图面的美学效果在显著增强，可读性在进一步提高，设计的趣味性以及确定性都在不断增加，如果说前一阶段需要的只是结构的清晰准确，这一阶段需要的更多的是形式的多变和空间的趣味。同一色彩的退晕、渐变明显增多，构筑物的形式以及必备的功能区域如停车场，开始显现。

The left drawing is a local detailed drawing of Wuhan Moshan Scenic Zone on the previous page, revealing the enormous difference of drawing expression when the scale of drawing changes from 1:20000 to 1:2000. A line in the general plan has become a road, and a jungle, even a brush stroke of marker has been enlarged into a patch of agricultural garden. In this stage, all forms and spatial elements begin to play a role. The water reveals difference in color. Patches of land reveal different cold and warm color, and islands show various forms of connections and roads. In this stage, the aesthetic effect of the drawing becomes more prominent and the readability, as well as the enjoyment and certainty of the design, is further improved. If the previous stage only requires for the clear and accurate expression of structure, this stage requires more the variety of form and the enjoyment of space. The gradual change of the same color significantly increases. The form and the necessary functional area, i.e. the parking lot, of the structure begin to emerge.

磨山风景区农业园景观规划
Landscape planning of agricultural garden in Moshan Scenic Spot

　　这是唐山南湖在水生态治理工作基本完成以后所建的第一个供市民使用的公共项目规划（2013）。本区域在 2016 年唐山承办世界园林博览会时由中央市民广场改作主入口，空间发生了很大变化，本图所列更多的是从草图绘制的角度说明在中等尺度 1:1000 左右的比例尺下景观手绘图可以达到的细致程度和美学效果。

　　就整体表达而言，采用了深色的水、中性的树丛和浅色的光差、花田三个层次对比。就线条的表现而言，属于休闲和外围自然区域均采用柔美婉转的曲线，而属于公共空间的大型广场，则采用最直接的直线型道路，目的是形成贯通城市与湖区的最通透的视觉廊道和最直接的功能通道。同时，也使应急、急救以及领导参观的车辆可以在 1 分钟之内达到最核心的景区。设计者用手绘图的色彩、线条的缓急等手法表达了这样的构思和规划目标。

　　This is plan of the first public civic project (2013) in Tangshan after the fundamental completion of waste water ecological management of Southern Lake in Tangshan. Of course, the space of this area underwent great changes when the central civil square was transformed into the main entrance in 2015 when the International Horticultural Exposition was held in Tangshan. This drawing shows the delicacy and aesthetic effects that landscape hand drawings can achieve under medium measuring scale of about 1:1000 from the perspective of sketch making.

　　As for overall expression, the comparison of three layers is formed with dark water, neutral bush and light-colored light equation and flower field. As for the expression of lines, graceful and euphemistical curves are adopted for relaxation and periphery natural area, while the most direct straight-type roads are adopt for large square in the public space. The purpose is to form a most transparent visual gallery and the most direct functional gallery that connect the city and the Lake District. This design makes lots of visitor stay at the picturesque roads in the beautiful ecological garden. In the meanwhile, it enables the emergency and first-aid vehicles to reach the core scenic zone within one minute. The designer expresses such a conception and planning objective by means of the changing routes of the colorful liens in the sketch.

唐山南湖市民广场第一轮总体规划
The first draft of the overall planning of Southern Lake Civil Square in Tangshan

秦皇岛金梦海湾规划设计
Planning and design of Golden Dream Bay in Qinhuangdao

　　这是一张2013年的二草平面图，本意是在此基础上完成电脑渲染，再交付业主单位。但实际过程中鉴于画面已有的效果，并未使用电脑，而是直接将工作草图作为成果交付业主，并得到了高度认可。这张图的创作过程说明一个问题，工作草图在图面表达效果尚佳的前提下，完全可以作为设计成品交付使用，目前规划设计界以电脑图纸作为唯一工作成果的认识是不全面的。

　　画面效果而言，作者刻意使用了同类色递进的方式及从城市第一层边界的灰绿色（设计边界以外）到鲜艳的紫色花田、蓝色绿道、棕色的海滨灌草区域、土黄色的沙滩、柠檬黄的前部海岸、淡绿色的水线和深蓝色的海洋，整个画面似乎是一个三棱镜折射出的七彩光柱，但其中蕴含的设计意图是用不同的色块将区域的功能明显地划分出来。目的是让业主在解读画面的第一分钟就全部读懂其中的设计信息。就像我们对悉尼歌剧院，对阿拉伯帆船酒店的认识一样。伍重的贝壳设计最大的好处是让人在看到作品的第一秒钟就能够确认这个建筑绝不会与世界上任何一座其他建筑相混淆。这是设计者真正的成功之处。

    This is a second draft plan made in 2013. The original intention is to render it on the computer on the basis of this sketch and then deliver it to the client. But in the real process, in view of the actual effect of the drawing, it was not rendered on the computer. The sketch was delivered to the client directly as the final drawing and was highly regarded by the client. The creation process of this drawing made a point that if the drawing effect of a sketch is well enough, it is fully acceptable to deliver it as the finished design product. The current cognition of the planning design industry of taking digital drawing as the only acceptable design result is not well-grounded.

    As for the graphic effect, the designer deliberately adopts the means of the progression of similar colors, which expends from the grayish-green boundary of the first layer (outside the design boundary) to the bright-colored violet flower field, the blue greenway, the brown coastal shrub area, the yellowish-brown sand, the lemon yellow front coast, the light-green waterline and the dark blue ocean. The whole image looks like a colorful light beam reflected from a triangle prism, but the implied design intention is to distinctly divide the functions of the areas with different color blocks. The purpose is to allow the client to understand all the design information in the first minute that he reads the drawing. It is just like our understanding of the Sydney Opera House and the Burj AL Arab. The biggest advantage of Jorn Utzon's shell-shaped design lies in that it ensures that people would not confuse this building with any other building in the world at the first second they see the work. This is the designer's true success.

以唐山 2015 世界园林博览会总体平面为例，同样这也是一张以完全的手绘平面替代电脑图作为工作成果提交的最"出格"的案例，规划的主体山水形式由孟兆祯院士主导，作者只是根据其规划意图做出一张可以与大师作品相配的平面表现。整个方案用六张 A0 图纸拼贴而成，虽然面积大到 500 公顷以上，但许多分景园的设计依然具有可读性，甚至具有一定欣赏性。故作者将其中有代表性的表现技法分块截出，呈现给读者。

Taking the general plan of the 2016 Tangshan International Horticultural Exposition as an example, it is also a most "outrageous" example of a plan totally drawn with hand submitted as work result in place of digital drawings. The main landscape form of the planning is led by Mr. Meng Zhaozhen, an academician. The designer made a drawing with a plan expression that is comparable to the works of great masters based on its planning intention. The whole scheme is collaged with six A0 size drawings. Although the total area surpasses 500hm2, the design of many split landscape garden still have readability and even a certain degree of appreciation. Therefore, the author selects parts of the drawing that reveal representative expression methods and presents to the reader.

唐山 2016 世界园林博览会场地总体平面图
Overall plan of the site of 2016 Tangshan International Horticultural Exposition

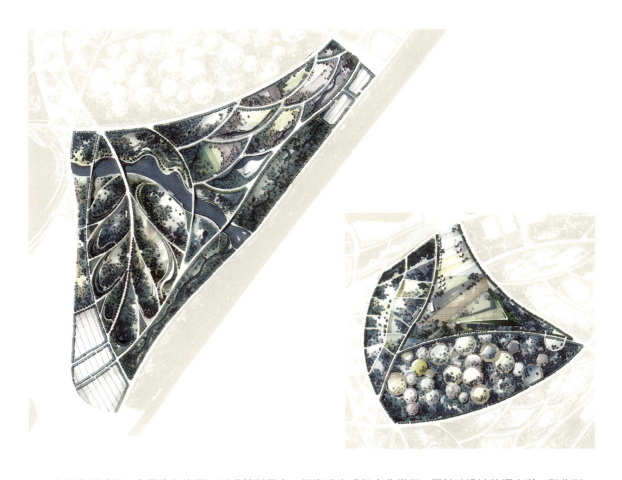

　　左图为树叶园，右图为气泡园，形式差别很大，但构造方式却十分类似，用基础设计的语言讲，即分形。这里边包括树叶叶脉的分形、气泡的分形，每一个分形单元里面，其实就是一座1000～2000平方米的小型园林，其基本结构都是外围的大乔木，立面的空间留给未来的设计师做出个人作品，可以说，构建世园会也是在搭建一个舞台。一个总体的公共空间舞台，和100多个小型的半封闭的各种风格造园家的舞台。但是，作为世园会的整体设计是完全不可能在一张二维平面中表达出100多种风格，否则，画面一定会混乱不堪，但反之留出100多个空白对设计者而言同样是一种失败，所以作者充分利用了手工草图的模糊性和不确定性，用一色的灰度的丰富变化填充了100多座分景园，使之既丰富又不确定，最终形成的是一个对于世园会整体场地的整体认识，其中确定无疑的只有路网和外围丛林的结构。

　　The left drawing shows a leaf garden and the right drawing shows a bubble garden. Although their forms are quite different, their construction modes are very similar. If said with the language of basic design major, it is a fractal problem. It includes the fractal of leaf vein and the fractal of bubble. Each fractal unit actually refers to a small garden with an area of 1000~2000m². Its basic structure keeps peace with the peripheral megaphanerophyte, and the elevation space is left for the future designer to make personal creation. Therefore, to a certain extent, the construction process of the International Horticultural Exposition is also the process of building a stage. It includes an overall stage in the public space and more than 100 small and semi-enclosed stages of various styles for the garden designer. But, as the overall design of the International Horticultural Exposition, it is totally impossible to show more than 100 styles in a two-dimensional plan. If so, the image would look messy and chaotic; whereas, leaving out more than 100 blank spaces is also a failure for the designer. Therefore, the designer takes full advantage of the fuzziness and uncertainty of sketch and fills more than 100 split landscape gardens with the changeable grey levels of the same color, making it both rich and uncertain. Finally, an overall understanding of the overall site of the International Horticultural Exposition is formed, in which the only certain part is the structure of the road network and the periphery jungle.

本图体现了中心场地的煤矸石山，经过两年改造以后的状况，通过统一的覆土和简单的栽植，一座白色的污染山已经被改造成了初步具有种植力的浅绿色山体（灌草种植），作者的任务是给山体披上一个具有形式感的丛林外衣，作者在山顶上设计了直接指向中央三角形区域的山顶平台，以此为核心设置了一系列螺旋放射状的树阵。表达出一个具有向心性和统领作用的中心景观。山下采用了柳叶型为主体的大片湿地，所有的道路、栽植完全依据这个形态做放射形的渐变，包括形态大小和色彩两方面的渐变。所有的树变成了空间中的一个点，用色的深浅完全依据外面结构的需要，真正的骨架变成了路网，整体色调只分两个层次，即水面和陆地。同样能在局部体现出整体的清晰与可读性。

This drawing shows the current condition of the gangue mountain in the central site after two years' transformation. With unified earthing and simple plants, a white polluted mountain has been transformed to a light green mountain (shrub planting) with preliminary capability for planting. The task of the designer is to put on a form of jungle coat for it. The designer designs a peak platform directly pointing to the central triangle area on the mountaintop, centered on which a series of spiral and radial modular planting is set. It shows a central landscape with centrality and leading function. Large patches of willow-leaf-type wetlands are adopted at the foot of the mountain. The radial gradual change of all roads and plants is designed entirely based on this form, including the gradual change in two aspects of morphology size and color. All trees become a spot in the space, and the darkness of the coloring is totally according to the requirement of external structure. The true framework becomes road network and the overall tone only has two layers, namely the water and the land. It can also reveal the overall clarity and readability in local.

景观规划设计平面
Plan of landscape design

上图是这一区域的放大，同样在微观层面（1公顷左右）在十余个分景园之间取得大小和结构的完美平衡，这里的平衡包括在色彩上由冷绿色到暖黄色，比例尺大小从1:5000平方米到1:1000平方米的梯度变化。用更为艺术的语言表达就是音乐的花园，"do-re-mi-fa-so……"，大小和色彩冷暖变化体现出的节奏感和音乐的道理完全相同。

下图为另一区域的分景园，整体用色略有变化但仍然符合这样一种渐变与层次对比的规律，无论表现形式是圆形还是菱形，这里明与暗、冷与暖、刚与柔、直接与曲折都成为画面丰富性表现的手段，总之初学者理解的丰富性往往是用色多而对比明确直至杂乱，而真正意义上的丰富性是在此基础上微妙而有节制的变化。记住，无论你花多少时间完成一幅作品，但对于观赏者而言，只会用1~2秒钟鉴别作品，如果你的作品在一两秒之内被评判为杂乱和费解，这就是设计者和表现者的最大失败。所以作者一直在强调整体。均衡之间的微妙变化。也就是说，即便用了一百种颜色，传递给读者的最好只有三个明确无误的层次。尽管用10小时完成这幅作品，观者只有两秒钟鉴别成败，这是我在平面草图阶段想传递给读者的信息。

The above drawing is a enlarged drawing of this area, which also achieves a perfect balance between size and structure among tens of split landscape gardens at micro level (around 1 hactare). Such a balance includes the change from cold green to warm yellow in color as well as the change of gradient from 1000 squaremeters to 5000 squaremeters. If expressed with more artistic language, it is a musical garden with the rhythm of do-re-mi-fa-so…..; the principle of music is exactly the same as that of the rhythm revealed by the change of size and color tone.

The below drawing is a drawing of a split landscape garden in another area. Although the overall coloring is slightly different, the compassion of gradual change and gradation still accords with the above-mentioned rule. Regardless of whether the expression form is round or rhombus, the light & dark, cold & warm, firm & gentle and straight & circuitous presented here all serve as means to enrich the expression of the drawing. In short, beginners always think that richness refers to the variety and comparison of colors so that they sometimes make the drawing look messy. However, richness, in its true sense, refers to the delicate and restrained change on this basis. Remember, no matter how much time you spend finishing a work, the audience would only take 1~2 seconds to judge it. If your work is decided as messy and unintelligible, it would be the biggest failure of the designer and the presenter. The integrity and subtle change with balance is what the author always emphasizes. That is to say, even if you have used 100 colors, it is better that only three definite layers are delivered to the reader. The message that I want to send to the reader in the plan sketch stage is that even if it takes you ten hours to finish a work, it only takes the audience two seconds to decide the success or failure of the work. This is what I want readers to know in the period of sketch.

度假区的规划，采用了较为轻松的表现方法，重点表现出南、北、中三个区块，人工建筑，绿地和水面的对比，在核心区与度假区之间设置了较大的绿带，在东侧的唐曹高速公路与场地之间设置了多个入口，将场地的可达性发挥到最大化，同时与中部生态核心区区分开来。具体的表现方式用深浅各异的绿色表现生态基底，红色的人工构筑作为点缀，但同样整体画面还是保持在一个灰度模式当中，有一个基本经验，读者一试便知，在灰绿色的底色下，即便是一层冷紫色，只要稍稍有一点红色因素，都会给人红色块的感觉，所以作者所指的画面的红与绿绝不可简单理解为高纯度的红绿，果真如此画面会相当扎眼、相当俗气。南区规划形式没有完全确定之前在图面表达上采用了概念性的灰白色体块，只是区分了建筑范围和主要的交通干道及绿道，其余信息均留在下一轮详细图纸中确认。所以画面南北两个地块在表现程度上分属不同的两个层次，这也是在设计过程中经常遇到的情况。

　　The planning of the resort adopts a relatively relaxed expression technique with emphasis on the expression of three blocks in south, north and middle and the comparison among the artificial building, the greenbelt and the water. A large greenbelt is set between the core area and the resort. Multiple entrances are set between the Tangshan-Caofeidian Expressway in the east side and the site, maximizing the accessibility of the site and distinguishing it from the central ecological core area. The specific expression mode is to use different greens to show the ecological base and red artificial structures as embellishment, but the overall image still stays in the same mode of grey level. There is an experience that readers can try out: with grayish green as the background color, even a slight amount of red element on a layer of cold violet would give people the feeling of red chunk. Therefore, the red and green of the image referred by the author cannot be interpreted simply as highly purified green and red; otherwise, the image would look rather dazzling and cheesy. Conceptual gray blocks are adopted for drawing expression before the planning form of the southern region is finally settled. It only distinguishes the scope of construction, major traffic trunk road and greenway. Other information is left to be confirmed in the next round of detailed drawings. Therefore, the expression degree of the south plot and the north plot of the drawing belong to two layers, which is a frequently encountered situation in the design process.

南部度假区总图，这是一张更为清晰的局部图纸，画面上半部分的建筑细节已经直接表现到了独栋别墅的层面，画面依然只有红绿两个色调，即用不同灰度的绿去衬托不同灰度的红，画面越接近核心区域，色彩越暖，外围区域一律采用冷色，由北到南的画面表达与功能完全吻合。北区一律由停车场围和，从主干道引入四条支路，形成适度便捷的可达性。向南为大容量的核心度假景区，中央湖区的低密度别墅区（分时度假产业）。再往南即新城镇开发，画面通过明确的色块分出了缓冲区、核心区以及性质完全不同的核心区，三类用色在主要的交通动脉和中央生态核心区之间设置了宽达数百米的大绿量的廊道。形成清晰的从上至下、从左至右的规划结构。图式语言有时能够表达出文字所无法描述的复杂情景。第一层次的规划中对于黑、白、灰，方圆曲直、轻重缓急等形式语言的合理运用，将有可能将一张晦涩的规划图纸变得结构清晰明了。这种表达更多的是意识层面，而非时间限制，经验丰富的设计师可以在作图的第一笔就开始灌注早已了然于心的景观结构，而多数的在读学生和初期的景观设计师，在此类项目中探究问题时，也更多地将各种混乱归结到形式方面的问题，这是需要有意调整和纠正的。多做大型项目，多做手图色块的对比，在一草与二草之间经常做交互性的工作都能极大提升自己对于结构的清晰的预见能力及表现能力。

This is a clearer local drawing of the general plan of the southern resort. The construction details of the upper part of the drawing has directly revealed to the level of individual villa. There are still two color tones of red and green in the drawing. Greens of different grey level are used to set off reds of different grey level. As it comes closer to the core area, the color gets warmer. All periphery areas adopt cold color. The drawing expression from north to south is totally consistent with the function. The north region is balanced with parking lot. Four branches are brought in from the trunk road to form a moderate and convenient accessibility. Core holiday resort of large volume and villa district (timesharing industry) of low density in the central lake area locate in the south region. Moving further south is the development of new town. Definite color blocks are used to separate buffer area, core area and core area with totally different nature. Galleries of large green quantity with breadth of several hundred meters are set between the main traffic artery and the central ecological core area with three types of coloring. A clear planning structure from top to bottom and from left to right is formed. Graphic language sometimes is able to describe complex scenes that cannot be described by words. The proper use of graphic language, i.e. black, white, grey, square, round, curving and straight and routings of the lines, in the first layer of the planning makes it possible to clarify the structure of an obscure planning drawing. This kind of expression lies more in the conscious level and is not limited by time. Experienced designers can pour the landscape structure that has been quite familiar in his/her heart into the drawing at the first stroke of the drawing, while most students and inexperienced landscape designers blame various confusions to the problem of forms when discussing this type of problem. This is a point that requires for intentional adjustment and correction. The ability of a designer to foresee and express a clear structure can be tremendously improved with the experience of doing large-scale projects, conducting comparison of sketch color blocks and frequently carrying out interaction work between the first draft and second draft.

唐曹高速西侧休闲度假区规划
Planning of the leisure resort to the west of Tangshan-Caofeidian Expressway

1.3 景观规划平面概念表达
1.3 Concept expression of landscape planning plan

  概念性规划草图，是涉及巨大尺度规划时必经的阶段。这一阶段的图面，对于美感不作要求，表达的工具根据个人爱好可随意选择，从墨线笔、记号笔到彩铅、色粉、马克笔均可以轮换使用，充分利用各种工具之间的覆盖力差别表象多层次的效果，最大限度表现结构清晰。这种多层覆盖式的工作方式易导致最终抹成乌黑一片，缓和它的方法有多种，比如用多层半透明的草图纸分别构建规划的路网、绿网、水网和竖向、植物等层次，最后像 PS 叠加一样将成果统一体现。但也可以用更简单的方式，比如将所有的想法用各种灰度色彩的半透明的马克笔表现，在第二层次用具有一定覆盖力的记号笔修改，这一层越重要的信息就用纯度越高的色彩表达，比如路网用红色，水网用纯蓝色，绿网用纯绿色……最后对于关键信息比如主体建筑、中央干道等直接拿厚重的涂改液一次性画出。这样做虽然画面依然杂乱，但绝不至于混乱乃至错乱，传递出错误信息，这一阶段首在准确，对于画面效果不应过度要求。

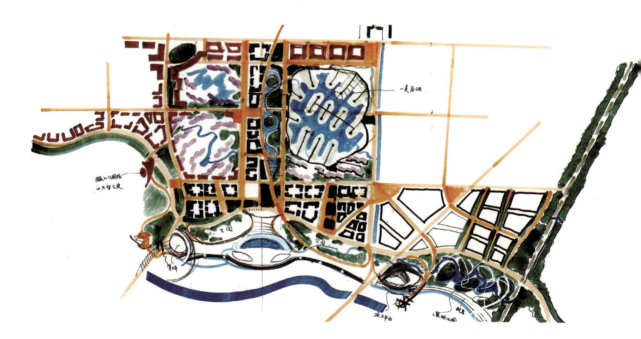

山西朔州七里河景观规划
Landscape planning of Qilihe District in Shuozhou, Shanxi Province

Conceptual planning sketch is an inevitable stage when it comes to the large-scale planning. There is no requirement on the aesthetics of the drawing in this stage, and the tools for expression can be chosen randomly according to the designer's preference. Taking full advantage of the different covering power of different tools, inked line pen, permanent marker, color pencil, color power and marker pen can be used in turns to express multi-layered effect and express a clear structure to the greatest extend. If such a multi-layered overlying method is adopted, the final effect would inevitably look jet-black. There are many ways to alleviate it. For example, multi-layered semi-transparent sketching paper can be used to construct different layers of the planning, i.e. the road network, green network, water network, elevation and plant, and the unified result can be shown in a similar way to the overlying effect of Photoshop. There are also simpler ways: all thoughts can be rendered with semi-transparent markers of various grey levels. Marker with a fair covering power shall be used for modification in the second layer. In this layer, the more important the information is, the purer the color for rendering is. For example, red is used to express the road network; pure blue is used to express the water network, pure green is used to express the green network; central artery is directly drawn with thick correction pen once. Although the drawing still looks messy with such a practice, but it would never go so far as to be chaotic or confusing to deliver wrong information. The primary concern in this stage is accuracy, and excessive requirements shall not be put on image effect.

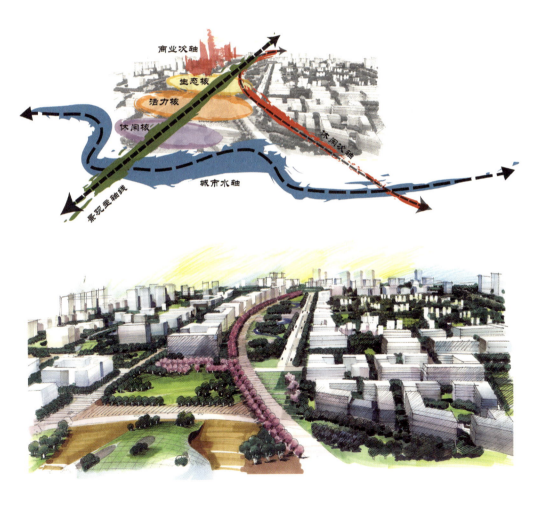

右下图是在整体效果图的基础之上用PS色块覆盖形成清晰的结构示意，右图则直接以景观结构为基础做出绿色网络图。整个图纸涵盖面积数百公顷，基本没有表现的内容，完全围绕景观结构的清晰性展开，自然的湖面用紫色、专门设计的水线河流用深蓝色、保护性绿块用深绿、休闲型绿块用浅绿……换言之，所有的色块都具有像文字一样的叙述表意，同时具有清晰的图面结构，这是说明性图纸必须重视的。清晰远比美观更重要。为了进一步清晰，其实还可以采用多层叠加的方式。总之，对于大型规划项目的初期规划而言，任何图纸都应该成为清晰的思维结构体的一部分，这一阶段更需要的是科学家、哲学家的思维，而不是画家的天马行空。

The drawing on the right bottom forms a clear representation of structure on the basis of the overall effect drawing with the covering of color block via Photoshop, while the drawing on the right forms a green network plan on the basis of landscape structure. The whole drawing covers an area of hundreds of hectares. It basically has no expression contents. The whole drawing centers on the clarity of landscape structure. Violet is used to express the natural lake; dark blue is used to express the specially designed waterline and river; dark green is used to express protective green block; light green is used to express recreational green block. In other words, all color blocks are narrative and expressive like words. It has a clear drawing structure, which is very important for illustrative drawing. Clarity is more important than aesthetics. Actually, to make it clearer, the method of multi-layered overlying can be adopted. In short, for the primary planning of large-scale project, any drawing shall become a part of a clear thought structure. This stage requires more of the thoughts of a scientist and philosopher, instead of the creative and unrestrained thoughts of an artist.

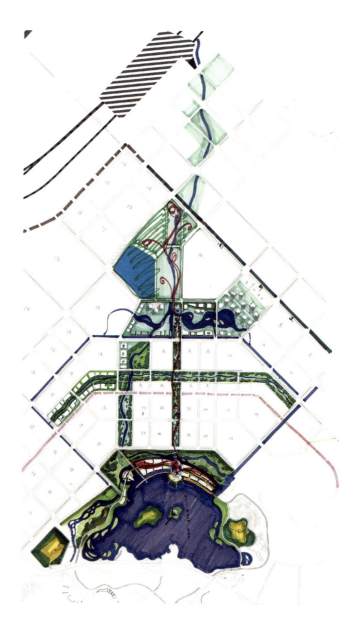

唐山南湖概念规划
Concept planning of Southern Lake, Tangshan

地块的简易划分：从简单的居住地块划分，到尽端路住区组团的设置，直至人车分离的专用休闲步道设置都可以用最简单直接的方式加以表示，做到功能划分明确，能够为下一步深化设计做好基础即可。

The simple division of plot: the simplest and most direct way of using A and B can be adopted for the division of simple residential plots, the setting of residential groups on cul-de-sacs and the setting of dedicated recreational footpath that completely separates pedestrians and vehicles. The drawing shall distinctively divide functional areas and lays a foundation for the detailed design in the next stage.

指状分布、尽端路系统，团块状居住地块
Finger-like distribution; cul-de-sacs system; clumpy residential plots

前页表现了规划起始阶段对于地块划分的几种最简单的表达方式，比如指状分布、尽端路系统、团块状等，表达的主体均为建设地块与交通、绿地之间的关系，在相同形状和相同尺度的地块内可以出现简繁不同的多个层次，比如第一排第一张与第二排第二张的区别，全部的六张图例从最简易的团块状地块分割，到道路系统串联的尽端路住区的组团划分，到微观层面两条高尔夫球道之间的别墅组团的划分，体现出这样一个道路，即便是在最枯燥的地块划分的规划平面中，也可以根据人的主观需求表达出不同层次的丰富性和趣味性。

本页上图是利用了 Google 平面和完全真实的场地鸟瞰，用单一蓝、绿、白色块重新勾画场地的景观结构。场地位于北京怀柔大沙坑地区。右图表现了沙坑现状与怀河之间的关系，左图鸟瞰则是将

The drawings on the proceeding page show several simplest expression modes for the plot division in the initial stage of planning, for example, finger-like distribution, cal-de-sacs system and clumpy shape. The subject of expression for all is the relation between the constructed plot and traffic and greenland. Multiple layers of different complexity can be shown in plots of the same shape and the same scale. Taking the difference between the drawing on right top and the drawing on middle bottom for example, such a road system is expressed in all six legends via the simplest division of clumpy plot, the division of cul-de-sacs residential group connected by road system and the division of villa group between two golf fairways in the micro level. Even in the planning of the most dull plot division, the richness and enjoyment of different layers can be expressed via the subjective demand of people.

The top drawing on this page takes advantage of Google plan and entirely lifelike site aerial view. Single blue, green and white color blocks are used to show

沙坑和怀河乃至远处的雁栖河、沙河结合为一个统一的景观综合体。单一的色块区分出了交通干线、绿廊、水脉，乃至两侧如画的稻田和远处的燕山等多个层次。将一张Google图像从高度的丰富复杂还原为基本的结构分析中所需要的简单清晰结构。

the outline of the site landscape structure. The site locates in the Grand Sandpit District of Huairou, Beijing. The right drawing reveals the relationship between the current sandpit and Huaihe River. The aerial view of the left drawing connects the sandpit, Huaihe River and the remote Yanqi River and Sand river into a unified landscape complex. Single color blocks are used to distinguish multiple layers from the traffic artery, pergola, water vein and to the picturesque rice field on both sides and the remote Yanshan Mountain. A Google image with high richness and complexity is reduced to simple and clear structure needed by basic structure analysis.

河北迁西滦河中央景观带规划总图
Master plan of central landscape zone of the Luanhe River in Qianxi County, Hebei Province

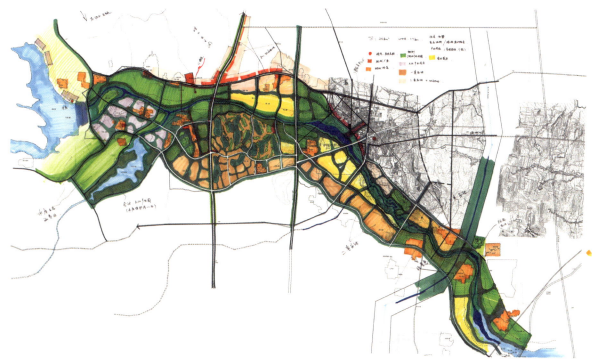

山西孝义胜溪湖湿地景观规划总图
Master plan of wetland landscape of Shengxi Lake in Xiaoyi County, Shanxi Province

  这是山西孝义一条河流与水库自然区域大面积场地的整体规划。在景观规划领域，与纯粹的城市规划真正的不同点就体现在我们以绿色基础设施的完整表达为目标，而非总规中的纯理性的文字导则，我们的规划总图同样与总规图有着明显区别。尽管在色块的区分上可以参照规划的法定色系，但在更具体的尺度上我们对蓝线、绿线以及发展区域的控制会比总规体现出更多更丰富合理的变通。这是在今后大规模的景观规划项目中所要坚持的本行业的特色，即从体制内的法规导则走向体制外的城市设计的中间阶段。本图即体现了这样的特色，即所有的建设性地块均模仿了总规图则的色系，但又体现出类似城市设计的微妙变化。在形式和法规两个层面上取得一种平衡。具体而言，在核心区地块的分布上，考虑的低冲击、低密度、低影响开发，外围随着色彩加深，项目的形式感逐渐加强，开始体现出城市在所有未涂色区域，保留了场地原有的城市化特征。这种渐变的画面效果是体现结构清晰的层次性的重要手段。

  This is the overall planning of a river and the large natural area of the reservoir in Xiao Yi, Shanxi province. In the field of landscape planning, the real differences from pure urban planning is reflected in our holistic expression of green infrastructure, rather than the purely rationality of the general rules. Our master plan is also clearly different from the general rules picture. Although the distinction between the color-block can be in accordance with the color regulated scheme in the planning, in a more specific scale, the control of the blue line, green line and the development region can reflect more abundant and reasonable changes than the general rule. This is the characteristics of the industry to be insisted on for large-scale landscape planning projects in the future. That is, from the rules and regulations of in-system to the outside system urban design of middle stage. This figure reflects the characteristics of all the constructive plots. They all imitated the color scheme of the general rules picture but reflect the subtle changes of similar urban design. In the both levels form and regulations to achieve a balance. To be specific, in the distribution of the core area, the consideration shall be given to low impact, low density and low development impact. With the deepening of color for peripheral area, the form of the project gradually escalates. It began to reflect the city in all uncoated areas. Retained the original urban characteristics of the site, such gradually change of the picture effect is an important method which reflects the clear hierarchy of the structure.

这是河北三河城市绿色系统规划图中的一部分，是典型的以形式阐述空间的图式。本图选用了规划设计中几个重要地段的局部图纸，表达一种从概念性规划到具体的形式设计之间的中间阶段。在画面所有元素中除了交通线和河流蓝线具有联系性以外，其余的场地都体现出块状、点状的特征，场地的具体细节在这个层次的表现中可以不予考虑，所有的植物都体现出围和空间的特征。重要的交通动脉和集中性的开阔场地，用暖棕色去体现，开阔的自然场地表现为淡黄色，连续的道路用红色和白色显著的标出，剩余的深色地块体现出绿色基础设施和结构围护体的特征。

以上所有图例均针对大型场地规划中经常遇到的问题提出了在设计表达阶段的相关技巧和策略。对于景观总体平面图而言，图中所有的色彩都不应被视为景观单元空间色彩的真实反映。同一片树林，在核心地段和在边缘地段可以体现出完全不同的两类色系，表达两种空间性质，天然水系和人工水系也可以用不同蓝色表现出来，建筑和交通线可以采用白色或红色，但最显著的对比色从空间中凸显出来，最终完成整体规划的骨架。整个的色彩配置均是依据设计目标的表现而设定，是典型的对客观物象的主观性表达。

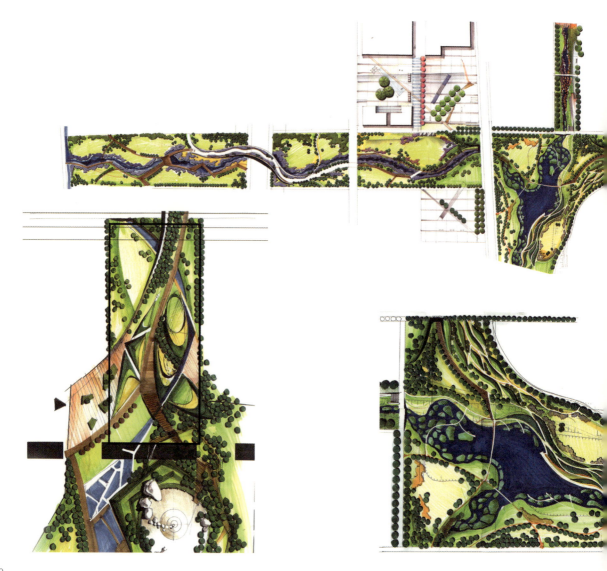

This is a part of the urban green system planning drawing of Sanhe City, Hebei Province. It is a typical schema that illustrates space with forms. Here are the partial drawings of several important sections in the planning and design, expressing an intermediate stage between conceptual planning to specific form design. Of all the elements in the drawing, only traffic line and blue river line show a connection with each other, and the rest of the site all present characteristics of blocks and points. The specific details of the site can be neglected in this expression level, and all plants shall present the characteristics of balanced space. Warm brown is used to express important traffic artery and concentrative open field. Light yellow is used to express open natural site. Consecutive roads are remarkably marked out with red and white, and the rest dark blocks shall show the characteristics of green infrastructure and structure enclosure.

All above legends propose relevant skills and strategies to frequently encountered questions during the planning of large-scale site in the design expression stage. For overall landscape plan, all colors in the drawing shall not be regarded as the true reflection of the spatial color of landscape unit. The color of the same patch of woods in core area and in edge area can be shown in two totally different color schemes to reveal two different kinds of spatial property. Different blues can be used to present natural water system and artificial water system. White or red can be used to present building and traffic lines. The most prominent contrasting colors are highlighted in the space to finally complete the framework of the overall planning. The overall color scheme is determined based on the expression of design objectives, which is a typical subjective expression of objective things.

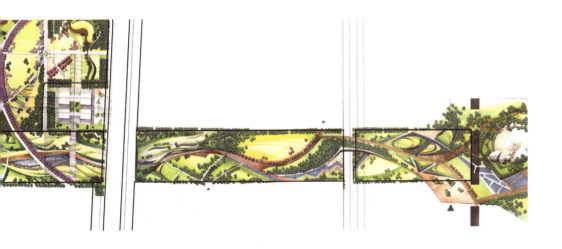

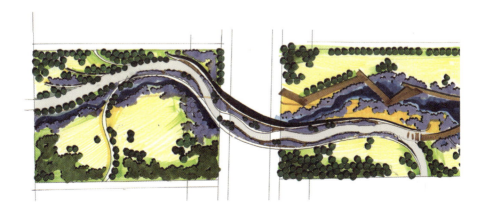

磨山风景区入口景观规划总图
Master plan of entrance landscape of Moshan Scenic Spot

第 2 章
剖、立面图的表达
Chapter II
Expression of Sections and Elevations

2.1 局部尺度的剖、立面图
2.1 Section and elevation of local scale

一个普通山谷的平面与剖面的对照，在平面图中很深的山谷通过剖面线的结合在剖面图上被完整体现出来，画面的所有色彩效果和笔触变化都体现在立面，所以剖面图不仅可以表现竖向，更可以通过透视形式的综合表达反映出层次和场景的总体氛围，换句话说，在自然风景名胜区一类的规划中，简易的剖面图只要稍加透视效果就可以表达出类似大鸟瞰图纸那样的丰富性。这一类图纸利用手工制作，其随意性、多选性和表达的效率都远远高于计算机建模和图像拼贴。后者在人工建筑密度较大的规律性空间中则展现出较多的优势。

The drawings reveal the compassion of the plan and the section of a valley. A valley that looks deep in the plan can be fully expressed on the section via section line. All color effects and change of brush strokes of the drawing are reflected on the section. Therefore, section can be used not only to express elevation, but also comprehensively reflect the hierarchy and overall atmosphere of the site via perspective format. In other words, in the planning of natural scenic spots, a simple section can become as rich as large aerial view in expression with a slight addition of perspective effect. This kind of drawing is suitable to be drawn by hand, and it's far better than computer modeling and image collage for its arbitrariness, compatibility for multiple choices and expression efficiency.

将剖面图和透视图的特征发挥到最大限度就能够出现如右图所示的效果，本图如按照剖面图的基本要求，可以被简化为一个平台支柱或一个前景大树等一根剖断线的单调构图，但是如果将这些元素细化并且按照透视原理表现，则图面出现极其丰富的表现效果。本图在以上图例所要求的清晰准确的基础上加入了更多对于细节的描绘，使一般清晰严谨的剖面图体现出如水彩画一般的质感和欣赏性。

The effect shown on the right drawing can be achieved if the characteristics of section and perspective drawing is developed to the greatest extend. As per the basic requirements of section, it can be simplified into a simple composition with section line, such as a platform support and a front view tree. But if these elements are detailed and expressed as per the perspective principle, the expression of the drawing would be able to present an extremely abundant effect. Under the premise of achieving a clear and accurate drawing, more details shall be depicted on the drawing, enabling an ordinary clear and precise section to show the texture and appreciative characteristics of watercolors.

本页及后页的三个剖面均是在极短的时间内为学生所做的现场示范图，体现了剖面图所需要的基本结构性元素，比如清晰肯定的剖面线，竖向的梯度配置，植物从乔、灌、草到水生花卉的过渡，左图偏重于景观的整体意象，中图是配景船的意向表达，本页上图体现了浅水湿地的基本构成，下图则对浅水区及排水设施做出了最简单清晰的表达。从生态手段到市政工程的结合均有所涉及，这种图式在团队协作阶段是最为快捷、清晰的结构表达方法。表现形式不求完善，可以是单一的黑白墨线图，也可以配合淡彩，总之，以表达出结构的清晰为第一目标。

The sections on this page and the opposite are all made in a short time while making demonstration to students, and they show the basic structural elements, including clear and definite sectional lines, vertical gradient configuration, and the transition of plants from tree, bush, grass to aquatic flower. The opposite drawing lays emphasis on the overall image of the landscape, and reflects the basic constitution of shallow water wetland. The left expresses the image of boat, and the right provides the simplest and clearest expression for the light water area and the drainage facility. This type of schema, involving the combination of ecological methods and municipal engineering, is the fastest and clearest structure expression method in the team cooperation stage. The pattern of expression can be a simple black and white inked line drawing, and it can also be cooperated with light color. In short, the primary goal is to express a clear structure.

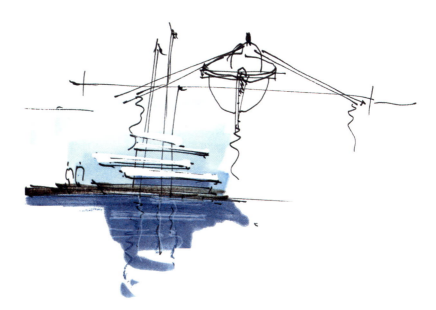

偏重于意象表达的景观剖面图

本页选择了三张课堂演示的图纸，表达了同一自然风景区三处具有特色的冲沟、峭壁和跌水景观，在背景表达上，都运用了出水最为充沛的 AD 马克笔，把油性马克的大色块和过渡性特征充分加以运用。山坡上的树木则大量运用了粗细不等的墨线，表现了结构的肯定，整体的色彩从天空的冷色到接近山崖的暖色形成明确的具有图示化效果的渐变。

景观的剖立面图主要反映标高的变化，地形的特征，高差地形处理及植物的种植特征。需注意：

（1）地形在立面和剖面中地形剖断线和轮廓线，要清晰明确。

（2）水面的水位线要表示清楚。

（3）树应当描绘出明确的树型，不同树种的绘制与配植，色彩变化与虚实对比。

Landscape section emphasizing on the image expression

Three drawings used for class presentation are selected on this page. There mainly focus on the image expression of the landscape profile drawing and express three characterized gully, cliff and waterfall from the same natural scenic zone. For the background expression, the most contrasting AD marker is used with full utilization of the large color blocks and transitional features of oil-based marker. Inked lines with different thickness are heavily used for the trees on the hillside to show a definite structure. The overall color from the cold color of the sky to the warm color near the cliffs forms a clear gradient with graphical effects.

The section drawing of the landscape reflects the elevation changes, terrain features, treatment of terrain with altitude difference and the planting characteristics of plants. It should be noted:

(1) The sectional lines and contour lines of landform on the section and profile shall be clearly and accurately depicted.

(2) The water line should be clearly and accurately depicted.

(3) The shape of the trees shall be clearly traced out, and attention shall be paid to the rendering of different varieties of trees and their complementary plants as well as the change of colors and the contrast between blurring and clarifying.

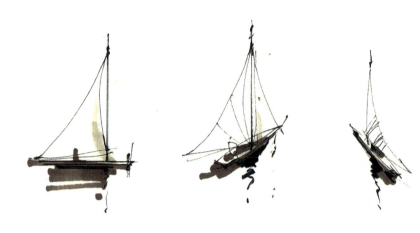

本图为秦皇岛海滨某俱乐部码头的一套完整的草图,从剖面到结构均作了清晰的表达。上图用单一的色块黑白的渐变表现出建筑和场地的基本环境特征,下图则通过平面和剖面的对照使整个项目设计意图一目了然。左下的总体平面表现了码头所在位置的周边环境状况。通过三张连续的图纸形成一个完整的叙事性的图示表达,具有连续解读的特征。

This is a complete set of sketches of a waterfront pier club in Qinhuangdao. A clear expression is made in aspects from the profile to the structure. The above picture shows the basic environmental features of buildings and grounds by the black and white gradation of single color lumps. The overall plan on the left bottom expresses the surrounding environment at the location of the wharf, while the drawing on the right bottom clarifies the design intent of the entire project via the contrast between the plane and section. A complete narrative graphical representation is formed via these three consecutive drawings, which has the characteristics of continuous interpretation.

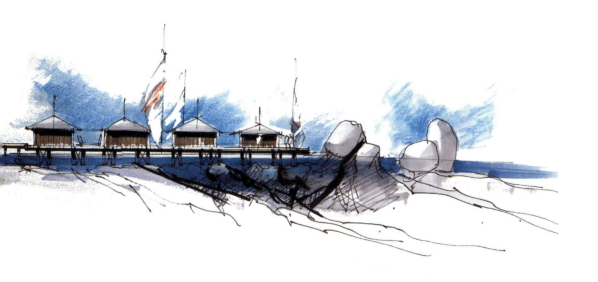

更为简易的剖面图纸，同样的山崖，可以用更简洁的笔法表现出最基本的信息。

This is a simpler section drawing. Simpler technique of drawing can be used to present the basic information of the same cliff.

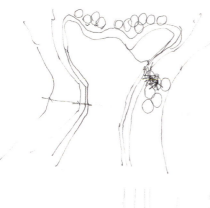

将剖面与透视的集中构图因素结合起来，形成更具有场景真实效果的同时又具有剖面图的准确性的综合表达。

With a combination of centralized composition elements of section drawing and perspective drawing, it forms a comprehensive expression that has both the real effect of the site and the accuracy of section drawing.

Pedestrian Entry.

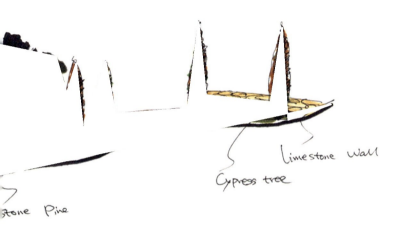

Limestone Wall
Cypress tree
Stone Pine

2.2 整体尺度的剖、立面图
2.2 Section and elevation of integral scale

剖面图：

（1）常需要扩张竖向，作非等比修改。

（2）剖面尤其应选择竖向较大的位置，剖面点和剖面方向选择很重要。

（3）剖视图极容易单调，因此应选择有标志意义的物象表现，如地域性植物、车、船等。

（4）剖面图可以很自然地变成透视图，甚至在同一图纸中表现也未尝不可。拉斐尔、伯拉孟特等文艺复兴大师均早就有过此类表达方式的尝试。

Section:

(1) It normally requires vertical expansion and non-geometric modification.

(2) Location with large vertical space shall be selected for the section drawing and the selection of section point and section direction is very important.

(3) Sectiondrawing tends to be monotonous; therefore, object with symbolic meaning shall be selected for more detailed expression, i.e. local plants, cars and boats.

(4) Section drawing can turn into perspective drawing naturally. It is even possible to show both on the same drawing, which has been tried out long before by Renaissance masters such as Raphael Sanzio and Donato Bramante.

具有更多随意性和表现性的剖面，左图重在表现树林的结构和植物的变化效果，右图主要表现湿地从灌木到草坡到湿地植物的梯度和植物变化。两张剖面的元素均极为简单，但由于加强了对树木、天空、水景的描绘，使图纸在表意的基础上增加了欣赏性和个性化特征。

These are more arbitrary and expressive section drawings. The opposite drawing emphases on the forest structure and plant change. The drawing of this page emphases on the change of gradient and plant from wetland to bush to grass slope to wetland plant. The elements on both section drawings are extremely simple, but thanks to the strengthened depiction of trees, sky and waterscape, the drawing is bestowed with more appreciation and personalized features on the basis of ideographical expression.

（1）风景大树分枝的变化，常规色彩变化。
（2）单株间的相互揖让，重合枝丫间的减省。
（3）前后层次，尤其是两株同种树之间的人为区分。
（4）辽阔山丘上的远景树，极为简约，却起着收住视线的画龙点睛的作用。

  条带状的植物配植在立面上注意远近景的叠透，前景树可以细致到结构、姿态、树枝、叶色的变化。远处成片的背景林则泛泛概括出体量颜色即可，远近景之间拉开距离。

(1) The change of branches of landscape trees and of regular colors.
(2) Trees shall avoid collision and the overlapping branches shall be deducted.
(3) The front and back gradation, especially the man-made distinction between two trees of the same variety.
(4) The remote view trees on the extensive hill looks extremely simple, but they serve the crucial effect of attracting the attention.

  The overlaying and penetration of remote and nearby view is a point to be emphasized on the elevation for the arrangement of stripped plants. The change of structure, posture, branches and leaf color can be meticulously depicted for the front view tree. For the remote patches of background forest, it is enough to only generally depict its volume and color. The remote view and the nearby view shall be distanced.

深圳平湖罗山自行车体育休闲公园山谷清音景区剖面
Section of Valley Voiceless Scenic Spot of Luoshan Bicycle Sports & Recreational Park in Pinghu, Shenzhen

这同样是一张将剖面图与多层次的透视结合到一起的综合性表达。从一个山坳去看整个山谷的景观，整体上体现了竖向的变化、坡度植物的梯度变化和景观栈道在前景的点缀，最终完成画面时补加了天空，整个透视图体现出一种类似效果图的复杂性和欣赏性。

This drawing also shows the comprehensive expression combining one section with multi-layered perspective drawing. It reveals the landscape of the whole valley from the level ground of one hill. It shows the overall vertical change, the gradient change of slope plant and the embellishment of landscape plank road in the front view. After everything is finished, the sky is added. The whole perspective drawing presents a kind of complexity and appreciation similar to working sketch.

深圳平湖罗山自行车体育休闲公园山谷清音景区剖面
Section of Valley Voiceless Scenic Spot of Luoshan Bicycle Sports & Recreational Park in Pinghu, Shenzhen

深圳平湖罗山自行车体育休闲公园山谷清音景区剖面图
Section of Valley Voiceless Scenic Spot of Luoshan Bicycle Sports & Recreational Park in Pinghu, Shenzhen

第 3 章
鸟瞰图的表达
Chapter III
Expression of Aerial View

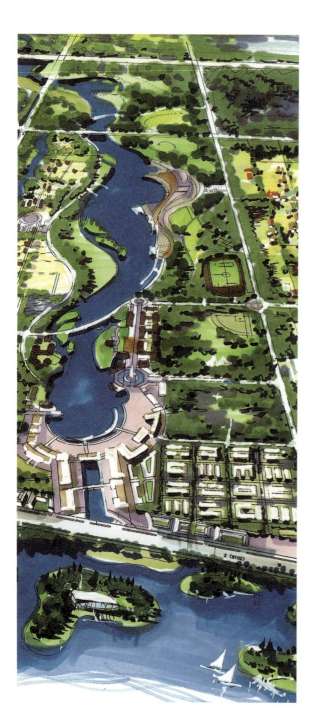

大尺度的鸟瞰，空间延展性大，草地、树群、林地往往需要用整齐有节奏的排线加以归拢，以达到空间元素的和谐统一，任何喧宾夺主的细节再精彩也是不必要的。

以下是油性马克笔配合彩铅的一种快速鸟瞰图表现。

景观园林重在气氛的营造，不论是写意还是半写意，不论是叶衬枝还是枝衬叶皆可，不论是哪一种方式，最重要的是画出厚度、体块，营造氛围。重在氛围、季相的表现，类似水墨写意。

Aerial view of large scale has a large spatial ductility. Grassland, cluster tree and woodland usually need to be linked together with trim and rhythmic rows of lines to achieve the harmony and unity of spatial elements. Any presumptuous detail, however splendid it is, would be deemed as unnecessary.

The following drawing shows a fast way to express aerial view with oil-based marker and color pencil. The key for the planning of landscape garden lies in the creation of atmosphere. Be it freehand brushwork or semi-freehand brushwork, no matter it is leaf setting off branch or branch setting off leave, whichever is the method, the most important thing is to draw out the thickness and block, and emphasis shall be laid on the expression of atmosphere and seasonal aspect, which is similar to ink freehand brushwork.

3.1 鸟瞰图步骤
3.1 Steps of aerial view

（1）线稿：起手宜松，做出大轮廓即可，局部可辅助排线光影，但不要画死，为着色留下余地。

（2）铺大色块：以季相、空间氛围描绘为主。

（3）细节描绘：加强近景,点出柏树阵。

(1) Sketch: the beginning lines should be loose. It is enough to only make the large outline. Local parts can be assisted with rows of lines and shadow, but it should not be fixed. Room shall be left for coloring.

(2) Pave large color blocks: it shall be centered on the depiction of seasonal aspect and space atmosphere.

(3) Depiction of details: nearby view shall be intensified, and cypress array shall be dotted out.

休闲农庄局部鸟瞰
Local aerial view of leisure farm

大面积的浅底色用"刷"，细节树群用点

large areas of light bottom color shall be "brushed out" and detailed tree cluster shall be dotted out

（1）用丰富的彩铅和色粉打底色。
（2）用润泽的油性马克笔"洗出"色底，形成细腻自然的退晕效果。
（3）用深色马克笔点出树丛暗部，最终形成类似水彩渲染效果。

(1) The bottom color shall be rendered with rich color pencil and toner.
(2) Color base shall be "washed out" with lenitive oil-based marker to form a smooth and natural retreating effect.
(3) Dark marker shall be used to dot out the shade of the bush in order to finally form a rendering effect similar to watercolor.

这是河流景观与城市关系的一张说明图，同时也是近景的河滨公园景观设计的效果表达，使用浓重的油性马克笔"点"出画面主体，形成类似水彩渲染的流动感和冷暖、深浅之间的退晕效果。

This is an explanatory drawing of the relationship between river landscape and the city. It also shows an effect expression of the landscape design of the riverside park with nearby view. Heavy oil-based marker is used to "dot out" the subject of the drawing, forming a retreating effect with the change of cold & warm and light & dark colors and a sense of flow similar to the rendering of watercolor.

南戴河中央公园鸟瞰
Aerial view of Nandaihe Central Park

深圳高铁北站"绿谷"大面积城市背景的做法
The practice of large scares of "green valley" city background for Shenzhen High-Speed North Station

（1）线稿为 Sketch Up 出图，稍加墨线整理，还要加上远景、地平线。

（2）铺大色块。天空一遍抹出，由黄到紫色，注意黄色部分较亮，为彩铅粗抹留下白纸形成，若加黄色马克，则会盖住宝贵的亮色；建筑主要在画影子和街道，建筑固有色其次，画建筑大面上要上浅下深，表现出阳光从上方投下的感觉；淡色抹出前景公园，一切至简。

（3）画出建筑细部及互相之间的投影，画楼群要有意分出组群，如此图前景中低层用暖色系稍加变化，后景高层用冷色系，前景多实体块（box），后景多反光材质，用以表现现代感的城市天际线。

(1) The sketch is made with Sketch up, added with the organization of inked line as well as distant view and skyline.

(2) Pave large patches of color block. The sky shall all be painted once from yellow to violet. The yellow part shall be bright to leave white bottom for the rough drawing of color pencil. If yellow marker is added, the valuable light color would be covered; for building, most attention shall be paid to the depiction of shadow and street, followed by the proper color of the building. The large surface of building shall be light on the top and dark on the bottom, presenting a feeling of sunlight casting from the top; the front view park shall be drawn with light color. Everything shall be as simple as possible.

(3) The architectural details and the projections between buildings shall be drawn out. When drawing building complex, groups shall be divided out with intention. For example, the middle-lower layers of the front view in the drawing are slightly changed with warm color scheme, and the high layers of the back view use cold color scheme. The front view is abundant in Box and the back view is abundant in reflective materials to show the modern urban skyline.

（4）补画出前景，调整画面。前景公园鸟瞰是作为城市配景，故虽处于前景，表达上仍以灰色为主，不宜太"抢"。前景的一组底层建筑以虚体表达，用灰色画出，目的都是为了突出中景的城市主题。

(4) The front view shall be added and the image shall be adjusted. Aerial view of the front view garden serves as the background of the city. Therefore, although it locates in the front view, the expression takes grey as the main color and should not be too eye-catching. A group of bottom buildings in the front view is expressed in virtual pattern and drawn with grey, with the purpose of highlighting the urban theme in the medium view.

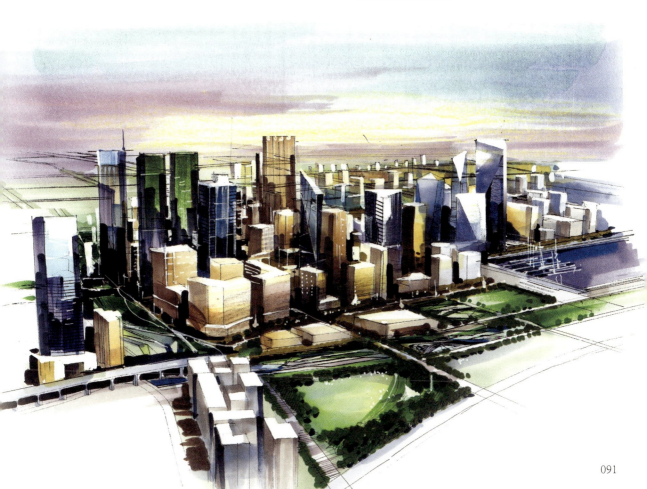

唐山南湖雕塑公园——纯粹自然为背景的鸟瞰园林

Southern Lake Sculpture Park in Tangshan——an aerial view of garden with pure nature background

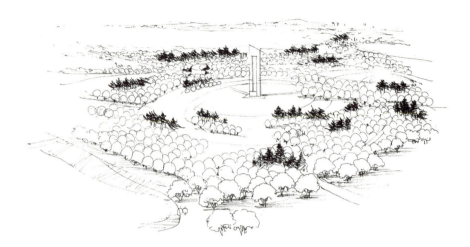

（1）这是一张纯粹以植物表现空间的场地，整体是一座山体绿化和一座小型室外剧场的表达。空间的表达要注意分出层次，分出群落以及常绿与阔叶。线稿要清晰，前繁后简，要避免画面画成"一锅馒头"，用钉头线做常绿树团，用笔要灵活。

（2）体色调平涂，要前紧后松，做出退晕层次。

（3）重点刻画前景树丛，从前到后色彩由暖到冷，由深到浅，由实到虚。冷暖色系交替要自然过渡，不可生硬。

(1) This is a drawing that purely uses the plant to express the space. The expression is an integral type of a mountain greening and a small size outdoor theater. The expression of space should to separate the sense of depth. Clusters, evergreen and broad leaf shall be separated out. The draft should be clear and complex in front and brief after. We should avoid drawing picture that confuse the primary with secondary. Nail-head line shall be used to represent evergreen bushes, and pen shall be flexible.

(2) Paint the overall tone in the flat manner, which shall be tight at first and loose afterward to make the gradation of changing colors.

(3) Focus shall be paid on the front view grove. From the front to back, the color should be from warm to cold, from deep to shallow, and from real to virtual. The alternation of cold and warm color scheme shall transit naturally without crudity.

（4）整理画面，须小心收拾，协调各部分大小的层次关系，补点人物和树木投影，完成画面。

(4) Organize the drawing with care. Coordinate the hierarchy of various parts with different size. The characters and shadow of the tree shall be added to complete the drawing.

唐山生态城广场——以城市为背景的公园鸟瞰
Tangshan Eco-Town Square——an aerial view of park with urban background

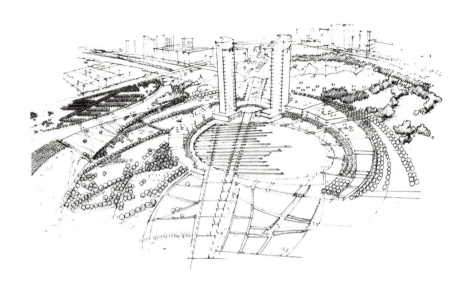

（1）利用Sketch Up导出的线稿进行处理，深化局部，更改（增加或删减，可先在Photoshop中进行处理再打印出底稿）模型前景变形的部分。

（2）快速平涂出天空及远景大色调。注意：1）退晕时底层可借用彩铅作底。2）画出大关系，尤其是体块间的阴影，只有阴影画到位，画中的物体、场景才能"落地"。否则，画面会"飘"。

（3）逐层刻画重点景物，中景采用最强烈的对比和最浓重的色彩。

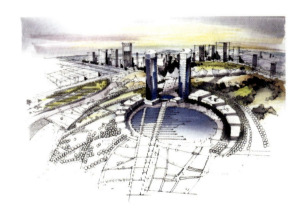

(1) The sketch exported from Sketch up shall be treated. Local parts shall be deepened, and the distorted part in the front view of the model shall be modified (add or delete; it can be pre-treated with Photoshop before being printed out).

(2) The major color tone of the sky and the distant view shall be painted out rapidly. Please notice: 1) to make color changing effect, color pencil can be used to render the bottom; 2) the major relationships shall be drawn out, especially the shadows between blocks. The object and scene on the drawing can only "fall to the ground" when the shadow is properly depicted. Otherwise, the image would look "floating".

(3) Key scenery shall be depicted layer after layer. The medium view shall adopt the strongest contrast and color.

（4）完成阶段最为关键，需调整画面前后关系，远景要退下去，用笔宜"虚"，前景用重色将场景层次区分开，最后点上人物、灯杆等细节。

(4) The completion stage is the most crucial stage. The front and back relationships of the image shall be adjusted, and the distant view shall be blurred. The brush stroke shall be "faint". The hierarchy of scenes shall be distinguished out with strong color in the front view, and the details like people and lamp post shall be added in the end.

第十二届园艺博览会·长沙总体鸟瞰
Overall aerial view of the 12th Horticultural Exposition, Changsha

(1)将大色块一次性"扫"出来,这一步通常可大量使用彩铅平涂退晕。空下画面的基调,天空可以在一开始就一次性画出,基调与远景视平线色彩一致。这一步也可将建筑的阴影和街道边界同时扫出来,这一步用笔要"整",不要抠细节。

(2)画出远景城市与主要建筑体块,亮部用建筑之间的投影衬出。

(3)确定阴影界面,调整色调,画出近景部分,坚持用大块笔触,整体色调要体现从远景浅色到近景深色的过渡。

(1) The large color blocks shall be "swept" out once and for all. This stage usually involves the flat painting and color changing effect made with color pencil. The keynote of the drawing shall be left out. The sky can be painted at the beginning once and for all, and the keynote shall be consist in color with the visual horizon of the distant view. The shadow and the street border can also be swept out in this step. The brush stroke in this step shall be "integral" without too much attention on details.

(2) Draw the distant view of the city and major construction blocks. The bright part shall be set off with the projections between buildings.

(3) Define the shadow interface; adjust the color tone; draw the nearby view part; stick to large-block brush stroke. The overall color tone shall reflect the transition from the light-colored distant view to the dark-colored nearby view.

（4）近景的细腻描绘，调整画面，完成。这一步整体色调的把握尤其重要，是"小心收拾"阶段，与第一步"大刀阔斧"的画风不同。

(4) Meticulously depict the nearby view and adjust the image to finalize it. The handling of the overall color tone is extremely important in this step, which shall be treated with caution and is different from the bold painting style in the first step.

山西孝义金龙山景观规划鸟瞰
Aerial view of landscape planning of Jinlong Mountain in Xiaoyi County, Shanxi Province

（1）以植物为主的景观鸟瞰，线稿部分注意整体构图，同时用笔应区分出植物的远近、常绿树、阔叶树等特征，远景山色只要画出轮廓即可。

（2）色彩的表现应突出季相，本图表现春景山杏、山桃盛开的寺庙景观，选用了紫色调，先大面积涂出天空和远山的退晕，用色要干净，远山与远处的天空要有所呼应，笔触表现要自然，天空最亮处可以用厚重的涂改液一次抹出。

（3）涂出近景色调，色彩要在冷紫色、浅蓝色和墨绿等色调之间相互交错变化，整体上寺庙的金色屋顶与山色的蓝紫色可以产生极好的对比效果，建筑用色要淡扫蛾眉，不宜来回涂抹，由于整体色彩对比强烈，金色的屋顶尤其不要过于浓重，建筑远景的基本规律是画出屋盖的投影，暗部用中性的灰色即可。这一步的远景可以用少许重色勾出暗处树丛的结构，同样要用笔寥寥，点到为止。

(1) This is a plant-centered landscape aerial view. For the sketch, attention shall be paid to the overall composition. The features of plant, including the distance, evergreen and broad leaf, shall be distinguished with pen. For the distant mountains, to draw out the outline is enough.

(2) Seasonal aspect shall be highlighted in the expression of colors. This drawing shows the temple landscape in the spring with the blooming wild apricot and wild peach. Violet color tone is selected. The sky and the distant mountain with changing color are painted out in large areas. The coloring shall be clear-cut, and the distant mountain and remote sky shall echo with each other. The brush stroke shall be natural and the lightest spot in the sky can be drawn with thick correction pen once and for all.

(3) Paint the color tone of nearby view. The color shall change among the cold violet, light blue and dark green, and intersect with each other. As a whole, the golden roof of the temple and the bluish violet of the mountain can produce a wonderful contrasting effect. The coloring of the building shall be slight and avoid repetitive back and forth painting. Due to the overall strong color contrast, the color of the golden roof shall not be too strong. The basic rule for the distant view of the building is to drawn out the projection of the roof, and neutral grey can be used for the dark part. In this step, the structure of bush in the dark in the distant view can be sketched out with a slight amount of strong color. The brush stroke shall also be concise.

（4）完成阶段，首先刻画出山顶建筑群的基本光影结构。由于环境色非常浓郁，主题部分反而可以以淡雅取胜，初学者在此往往会用力过猛，画过了，反而形成与前景色彩的冲突，本土的建筑结构应服从于光影表现，屋顶、台阶、墙面的受光面基本处理成白色，在绚丽的色彩背景下反而显得突出。近景补全要适当，近景的树丛面积大，变化多，稍微画过头就会使画面发花、显脏，所以，应点到为止，一般着力都在中景，远山虚，近山淡雅，才能使前后层次清晰，节奏明确。这就是所谓小心收拾。

(4) Completion stage: first, depict the basic shadow structure of the building complex on the mountaintop. Since the environmental color is quite strong, the subject part can use simple light colors instead. Beginners tend to overexert with excessive colors and form a conflict against the color of the front view. The building structure of this drawing subjects to the expression of shadow. The illuminated surfaces of roof, step and wall have basically been painted as white, which look prominent under the gorgeous and colorful background. The complement of nearby view shall be proper. Since the bush in nearby view is large in area and abundant in changes. Even slight overexertion would make the drawing look messy and dirty. Therefore, it shall be only to the point, and more focus shall be put on the medium view. Only when the distant mountain is blurry and the nearby mountain is light in color can a clear front and back hierarchy and a clear-cut rhythm be formed.

某城市滨水核心区规划意向
Planning intention of a certain urban water-front core area

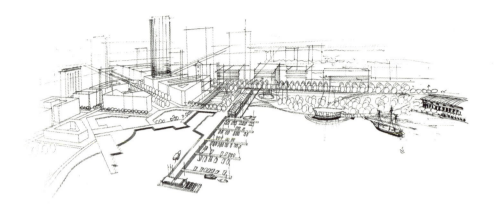

（1）利用 Sketch Up 模型导出简单的透视线稿，在此基础上用墨线画出细节。

（2）用灰色扫出建筑的明暗关系，并用墨线画出要素的阴影，使画面"立"起来。

（3）铺大的色彩关系，重点描绘滨水核心区的要素构成——木平台、滨水公园、码头等。

(1) Export simple perspective sketch from Sketch Up model, and draw details with inked lines on this basis.

(2) Sweep out the light and shadow relation of the buildings and draw the elemental shadows with inked lines to make the image look "solid".

(3) Pave large color relation and emphasis on the depiction of components——wood platform, Waterfront Park and wharf in the waterfront core area.

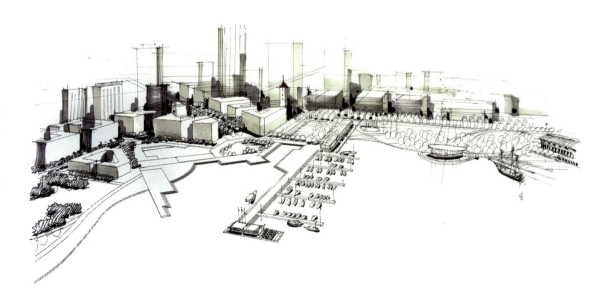

（4）完善画面，渲出水面和天空。细节元素的刻画是使画面动人的关键，比如楼群的质感、桅杆、游艇、海鸟等等。

(4) Complete the drawing with rendering of the water and sky. The depiction of detail elements is the key to make the drawing attractive, i.e., the texture of the building complex, the mast, yacht and seabirds.

河北曹妃甸生态城滨海大道景观鸟瞰过程图
Landscape aerial view process drawing of Caofeidian Eco-town Coastal Road in Hebei Province

（1）在简练的线稿上由远及近铺大色，远处的天空和水面颜色相同。

（2）铺场地色，远近可以有冷暖变化，远景缥缈，用冷色。城市部分用深绿色画出行道树进而衬出道路，用修正液点出中景和远景的楼群（修正液颜色稳定，盖多遍颜色并不会影响"高亮"效果）。

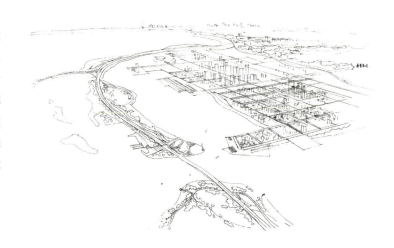

(1) Pave the colors of large areas from far to near on the succinct sketch. The color of the distant sky shall be the same as that of the water.

(2) Pave the color of the site with change of warm and cold color tones according to the distance. Since the distant view is misty, cold color is chosen. For the urban part, dark green is used to draw the border tree to further setting off the road. The building complexes in the medium view and distant view are pointed out with correction pen (the color of correction pen is stable. Multiple coverings of colors won't affect the "highlight" effect.)

（3）整理画面，用浅灰画出楼群投影，用高光笔勾勒出近景的园路和远景的桅杆，并将因快速上色盖住的部分跨海主路浅色重新"修补"出来。水面不宜全部涂色，只将倒影表示出即可，留白的水面是反射的天空的颜色。

(3) Organize the drawing: light grey shall be used to draw the projection of building complex. Highlight pen shall be used to sketch the contours of garden path in the nearby view and mast in the remote view. Partial sea-crossing main roads that are covered due to quick coloring shall be "repaired" with light color. The water shall not be colored entirely. It's enough to only express the inverted image, and the blank water is the color reflected from the sky.

3.2 鸟瞰图特殊表达技巧
3.2 Special expression techniques of aerial view

鸟瞰图综合画法中速度最快最直接的方式就是将现有场地的 Google 图纸直接下载并在上面用高覆盖力的色彩及涂改液进行再次描绘。本节以秦皇岛市中心的西海岸沙滩整体规划改造为例，利用 Google 的平面与鸟瞰用极快的后期改绘手法完成了沙滩周边的海港浴场、森林公园和别墅区域的多种表现。

The fastest and most direct way among the comprehensive techniques of drawing aerial view is to directly download the Google drawing of the current site and depict it again with color and correction pen of high covering power. In this section, the author takes the overall planning and renewal of Central West Coast Beach in Qinhuagndao City as an example and illustrates the method of re-depiction of plan and aerial view in later stage via Google, demonstrating various forms of expression for the harbor bath, forest park and villa area in the periphery of the beach.

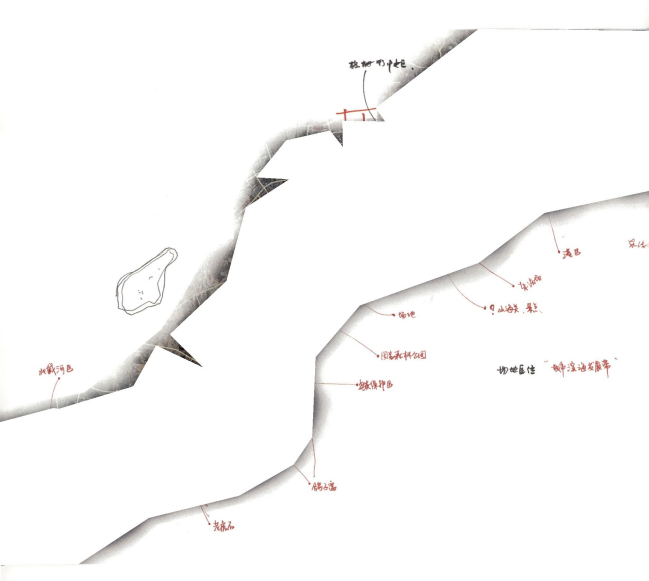

海水浴场西侧别墅区改造鸟瞰
Aerial view for the renewal of Villa Area to the west side of the Beach

用 Google 图纸作为底图表现大尺度景观的概念性方案是一种快捷有效而且表现力上佳的表现手段，实际操作过程中可以同时进行平面和透视鸟瞰的模拟，通常可以使用马克，彩铅和涂改液笔等具有一定覆盖力的工具，设计部分用彩色，高亮显示，Google 原图未覆盖部分则成为最真实的场地背景。

本图为秦皇岛西海岸四公里黄金海滩城市设计的真实项目。通过市域全图，场地局部和全景鸟瞰三张图纸将场地与秦皇岛城市，与南戴河森林公园以及一线海滩等场地要素完整真实地反映出来，同时用高亮度的涂改液直接画出场地的新建与设计部分，现状与设计两部分内容清晰可辨，更重要的是，这种方式极大加快了大尺度规划草图的设计进度，免去很多不必要的底图重复绘制过程。

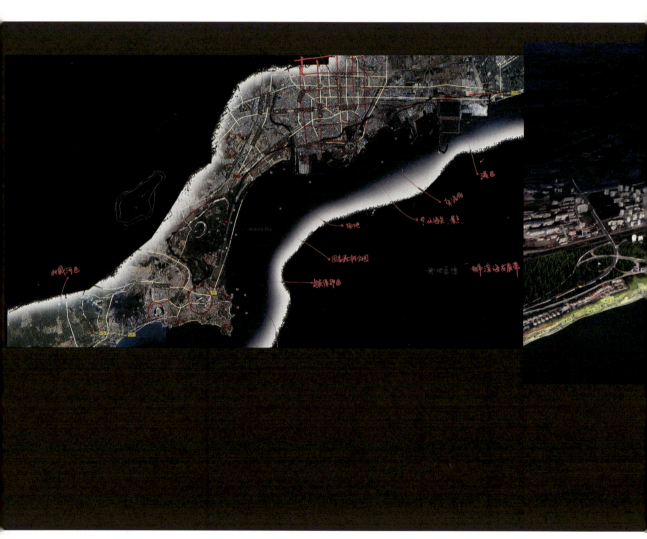

鸟瞰图画法——利用 google 做规划鸟瞰
Drawing of the aerial view-make the plan with aerial viewvia Google

Make aerial view plan with Google: To make conceptual scheme for large-scale landscape with Google drawing as traced drawing is an expression method that is fast, effective and expressive. In actual process, plan and perspective aerial view can be simulated in the same time. Normally, tools with a certain covering power, including marker, color pencil and correction pen, can be used. The designed part can be shown with color or highlight, while the uncovered part of the original Google drawing provides the most authentic site background.

This drawing is designed for a real project, golden beach of 4 kilometers in the west coast of Qinhuangdao city. Site elements, including Qinhuangdao City, Nandaihe Forest Park and Front beach, are revealed fully and authentically with three drawings, which are the general city plan, site local plan and panoramic aerial view. Correction pen with high brightness is used to drawn the newly-built and designed part on the site, making the current situation part and the designed part clear and distinguishable. What's more, such a method greatly accelerates the design process of large-scale sketch and spares many unnecessary and repetitive processes of making traced drawing.

场地总体鸟瞰
Overall Aerial Vie

根据 google 所做的港区和游艇码头草图。前景的游艇码头全部用涂改液直接在深色底图上绘制。场地现状的堆场和设计的新游船码头前后并列，所有的设计内容，包括码头、栈道游乐场、内港等元素全部用最简洁的方式绘出。前景黑白的设计主体与背景金色的沙滩和更远处蓝色的城市形成结构清晰、层次分明的设计整体。右图是在此基础上顺势做出的稍稍精细的表现图纸。

This is a sketch of harbor district and Yacht marina made based on Google map. The Yacht marina in the front view is directly drawn with correction pen on the dark traced drawing. The current storage yard on the site lies parallel to the designed new Yacht marina front and back. All the design contents, including elements of yacht, plank road amusement park and inner harbor, are depicted with the simplest way. The black and white design subject in the front view, the golden beach in the background and the remoter blue city form an integral design with clear structure and distinct hierarchy. The drawing on the opposite page is a slight refined expression drawing made on this basis.

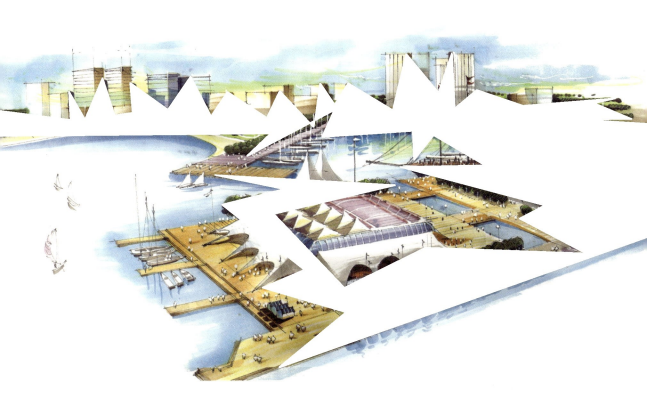

河北秦皇岛金梦海湾景观设计
Landscape design of Golden Dream Bay in Qinhuangdao

利用平面图扭转一定角度，可以很快形成一张透视效果图的底图，为其加上厚度，添置背景和一些景观细节，即可以很快完成一张准确的透视鸟瞰。对于以景观树丛、草坪为主的设计，这种方式可以做得非常快捷精确。唯一需要注意的是远近景的正确区分，对于一张经 PS 扭转角度的透视底图而言，远近景在精度上是没有区分的。在绘制透视表现时，需对前后景的表现稍作调整，前景有时需增加细节，而后景则需抹掉一部分过细的内容，如此才能跟人的欣赏习惯相符合。

场地总平面
General site plan

从总平面图中截取一部分作为设计表现的底图
A portion shall be cut out from the general plan and taken as the traced drawing of the design.

A traced drawing for perspective effect drawing can be formed rapidly by slightly rotating the plan with a certain angle. With thickness, background and some landscape details added on its top, an accurate perspective aerial view can be completed in no time. For a design centered on landscape bush and lawn, this method can be both fast and accurate. The only concern is the correct distinction of nearby view and remote view. For a perspective traced drawing formed by rotating angles via Photoshop, there is no distinction between nearby view and remote view in precision. When rendering the perspective effect, adjustment shall be done for the expression of front view and background. The front view shall be added with some details sometimes, while a part of too detailed contents shall be deleted from the background to be consistent with the appreciation habit of people.

河北秦皇岛金梦海湾景观设计
Landscape design of Golden Dream Bay in Qinghuangdao

直接在虚化的底图上做景观描绘，前景尽可能细致，后景则要适当精简，形成虚实层次变化
Landscape depiction can directly be conducted on the blurred traced drawing. The front view shall be as detailed as possible, while the background shall be properly simplified to form a change of gradation and contrast of blurring and clarifying.

山西孝义胜溪湖总图规划

通过各种类型的草图将尽可能多的设计信息图像化，包括交通现状，用地划分，水岸形态，空间视觉廊道均可用图式化形式分层次表现在一张或多张草图之上，还可以通过电脑图像叠加形成设计需要的多方面的说明图纸。

Master plan of Shengxi Lake in Xiaoyi County, Shanxi Province

The design information, including traffic situation, land division, shape of water bank and spatial visual gallery, shall be visualized as much as possible via different types of sketches. They can be expressed on one piece of several pieces of sketches by different layers with schematized form. All-round explanatory drawings needed for the design can also be formed via overlying of computer images.

山西孝义胜溪湖总体鸟瞰
Overall aerial view of Shengxi Lake in Xiaoyi County, Shanxi Province

第 4 章
景观多义表达
Chapter IV
Polysemous Expression of Landscape

景观在规划和策划阶段，利用手绘可以达到一图多义的效果，这就是手绘优于准确电脑制图的模糊性和多样可能性的表现。以下展示的是以一张总体平面图为基础，对各个局部多角度的空间模拟和再现，分别侧重于环境氛围、使用者感受、活动展开、植物季相等多种涉及诉求，所有的说明性图纸均做得小巧精致，往往能在十余分钟完成，极其快捷且说明性强。这种画法往往是草图阶段使用最多的表达方式，对于成熟的设计师而言，这种组图方式随意性强，表达的意象丰富，往往能信手拈来，更能以最流畅的手法表达出最初的设计灵感和思想。

这种方式对于学生在设计团队内部探讨以及对于入职、考研的设计表现，也同样具有极强的表意性。下图是作者专门为本专业考研学生所作范图，总平面表达为 1 小时，其余说明图纸均不超过 10 分钟。整个图纸可以在 2~3 小时内完成，大大低于目前景观专业研究生考试设计时限的要求，对于通常 4 小时的入职设计考试而言，这种制图方式也一样适用。

In the conceptual and scheming stage, the polysemy of drawing can be realized via hand drawing, which is a manifestation of how hand sketching is superior to computer mapping in its ambiguity and potential for diversity. This is also suitable for beginner to take part in the discussion among the design team, as well as in the beginning of job and the postgraduate entrance exam. Taking a master plan as the basis, all local parts are simulated and represented spatially from multiple angles, with respective emphasis laid on various design appeals, including environment ambience, user's feelings, activity development and seasonal aspect of plants. All explanatory drawings shall be small and exquisite. This painting technique is the most frequently used expression method in the sketching stage. For an experienced designer, such a method with drawing collections is highly arbitrary and able to express rich images. It can be drawn freely without too much hesitation, and the initial design inspiration and idea can be expressed via the most fluent means, which is extremely expressive.

唐山南湖休闲农场总图及意象

　　对页是唐山南湖中央公园湿地地区的一处局部表现，在高大的垃圾山下，以凤翎为主题（唐山是一座凤凰涅槃的城市）表现了一处兼具观赏性和生态净化功能的综合性湿地。从整体规划平面到一片凤翎形成的形式化的湿地，到湿地周边的椭圆形场地设计，到在此基础上人视点透视，形成一个景区完整系统的表达，具有连环画一般的叙事特征。本页下图也体现了同样的特征，在总平面上选取两处中心景观，简单勾画了景区鸟瞰，观赏者可以通过总平面的索引一眼辨识出场地的特征和容量范围，两张小鸟瞰图的总用时不超过 40 分钟。这种结合平面透视鸟瞰及细节的多层次表达是手绘设计当中一图多义，体现场地丰富性的重要手段。

总平面图
Master plan

Master plan and image of Southern Lake Leisure Farm, Tangshan

The drawing on the opposite page shows a local part of the wetland region of Tangshan South Lake Central Park. With the theme of phoenix feather (corresponding to the fact that Tangshan is a city that has undergone nirvana), it shows a comprehensive wetland under the towering garbage mountain that has ornamental values and ecological purification functions. From the overall planning plane to the formalized wetland formed by a piece of phoenix feather, to the oval site design at the periphery of the wetland, to the perspective of human added on such a basis, an expression of a complete scenic sport system has been formed, rendering it with the narrative features similar to comic strip. The drawing below on this page also shows the same characteristics. Two central landscapes are selected from the master plan and the aerial view of the scenic spot has been sketched in a simplified manner. The audience can identify the characteristics and capacity scope of the site at first sight with the help of the master plan index. The total time used to finish these two small aerial views is no more than 40 minutes. Such a multi-layered expression that combines plan, perspective, aerial view and details is an important means to present the polysemy of hand drawing design and show the richness of the site.

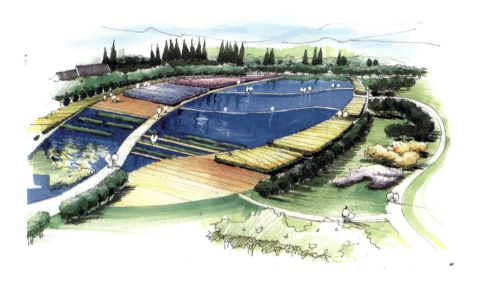

中心水花园鸟瞰
Bird's view of central lake

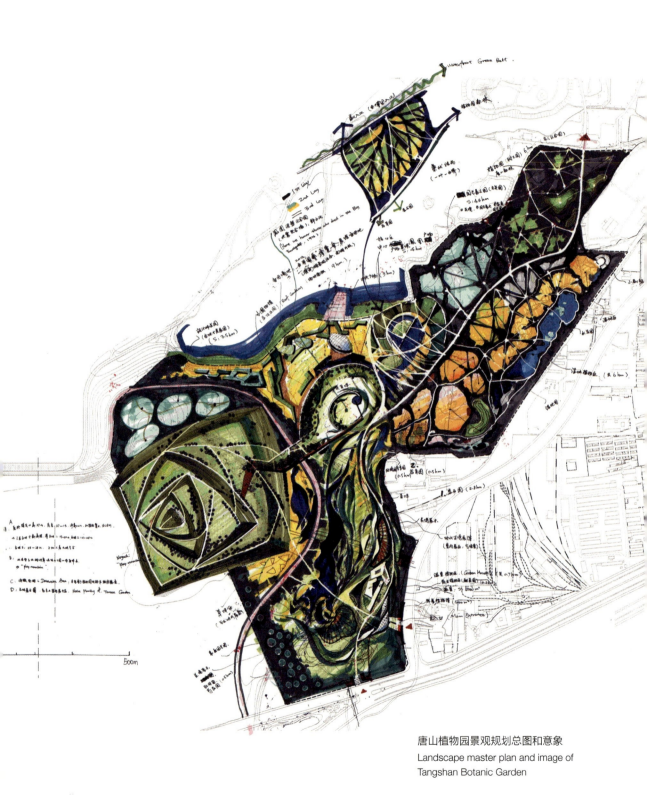

唐山植物园景观规划总图和意象
Landscape master plan and image of Tangshan Botanic Garden

唐山湾菩提岛规划设计

一个海岛的整体规划，对于介乎策划与规划阶段的草图表达而言，最重要的并不在于设计的精准性，而恰恰在于设计的多样性选择，并力求通过意象表达的方式提供给业主多种选择及其相应效果。本规划针对海岛的观光旅游和参与性活动为主的功能诉求，用概念图的方式表达了游艇住区，古迹区，渔人码头，小镇及海洋湿地探险区，并以空间营造和植物选择为重点，针对不同场景提供了功能展开和使用者角度的多个意象草图，极为方便快捷地表达出一个相当复杂的场地所要求的基本场景意象。

Design of Bodhi Island in Tangshan Bay

The sketch expression of the overall planning of an island. Between the stage of conceptual design and specific design, the most important thing does not lie in the accuracy of design, but in the diversified options of design. Multiple options and their corresponding effects are provided to the client via the method of image expression. Targeting at the major functional demands of the island, which is sightseeing tour and participatory activity, this planning shows the yacht residential area, historical sites, Fisherman's Wharf, small town and ocean wetland adventure area by the method of conceptual drawing. Multiple image sketches are provided for different scenes from the perspective of function development and users with the emphasis on the space construction and choice of plants. It is a convenient and fast way to show the basic scenario images required by a rather complicated site.

唐山湾菩提岛规划设计总图（500公顷）
Master plan and design of Bodhi island in Tangshan Bay (500 hectare)

对唐山湾菩提岛的多种小透视整体表达示例

Demonstration on various integrative expressions via small perspective with the example of Bodhi Island in Tangshan Bay

山东潍坊白浪河滨水景观带规划总图及意象

　　由日本某建筑事务所主持规划的山东白浪河入海口河段的城市设计，在其基础上的绿色系统规划表现当中加入了大量沿途景观构筑的小鸟瞰和空间意向图纸，将一张本来相当枯燥的难以展开的线状空间城市设计变成了清晰、丰富、可读的连环画式的作品。其中可以推敲的方面非常多，比如一座大型构筑物从公园绿地的内部观看，或从河对面观看的完全不同的景观意象，一座大众浴场白天所体现出的结构特征，和晚间灯光璀璨下的丰富意象都可以形成极为丰富的景观环境表达，同时在这些丰富的外表之下所透视出的，是对大众景观的民主意识的再诠释。这里既包含着类似刘易斯·芒福德所说的街头芭蕾的某些情感因素，也包括了我们在城市绿色基础设施建设完善过程中注入了具有中国特色的民主民生观念和平等利用的思想。这都是超于简单空间营造应有的思维方式。正如美国第一座具有大众平等思想和引导城市生活作用的纽约中央公园的营造，应该由一个新闻和社会工作者主持，在今天中国新城镇规划的社会历程中，景观师想要做的简而言之就是超越空间、构建舞台，建筑师的丰碑应该熔铸贯穿在所有大众舞台的营造之中，这是我们所谓同在一片屋檐下，共同使用一座城市以及让芭蕾回到我们的街头，这是一切呼唤的根源所在。

Master plan and image of waterfront landscape belt of White Wave River in Weifang, Shandong Province

The drawing shows the city design planning of the estuary of Shandong Whitewave River led by a Japanese architectural firm. Abundant small aerial views of landscape structures along the river and spatial intention drawings are added into the expression of the green system planning, which is carried out on the basis of the above-mentioned drawing. The drawing for the linear spatial city design which is originally boring and difficult to develop has been turned into a clear, rich and readable work similar to comic strip. There are plenty of aspects in the drawing that deserve to be discussed. For example, the landscape images of a large-scale building inspected from the interior of garden greenland would look totally different from that inspected from the opposite back of the river. The structure characteristics presented by a public outdoor baths seaside resort in the daytime and the rich images revealed under the splendid lamplight at night could both form extremely luxuriant expression of landscape environment. In the meanwhile, it is the reinterpretation of democratic awareness towards public landscape that is reflected out from these abundant appearances. It encompasses not just some emotional factors street ballet contains, similar to the words of Lewis Mumford, but also the idea of democracy and people's livelihood bearing Chinese characteristics and equal usage, which has been instilled into the process of improving urban green infrastructure construction. Both are beyond the due thinking mode of simple space construction. For instance, New York Central Park, which is the first to have popular equality idea inside and to guide urban life, is better chaired by a pressman and social worker. In today's new town planning process prevailing in China, a landscape architect's vision, in short, is to construct stage beyond space scope. Architects need to erect their monument in the whole process of building public stage. This is virtually the so-called using a city under the same roof and bringing ballet back to streets; this is the root of all calls.

北京永定河门城湖区景观规划总图及意象

　　本图将河流规划中最重要的水位及竖向设计信息作为表现重点，并在立面上加入了对于未来植物规划的一些建议，由此将一张本来纯属理性、科学层面的略显单调的图纸，通过透视图的引入变得丰富和具有欣赏性。当然，图中所有的断面效果都是依据场地的实际情况，都对场地的总体规划具有不可或缺的功能。离开这样一个前提，单纯的美就失去了意义。

　　上页及本页图纸属于同一项目的两个阶段，前者为偏重生态和水利设计的景观规划，包含了对于河床最佳流线的修复，多种水位适应性河床的营造以及相应的多层游步道设置和与水位相应的梯度植物配置，规划采用了多角度的剖面设计，分别展示了上部河床、下部河床及相应的植物配置，通过最简单的图例形式表达出复杂的设计内涵和意象。

Landscape master plan and image of Mencheng Lake District in Yongding River District, Beijing

This drawing takes the most important water level and vertical design information of river planning as the key point of expression, and give some suggestion to the future planting design on the vertical face. Thus, a monotonous drawing in purely rational and scientific level has become rich and appreciative with the introduction of perspective drawing.

The previous and the current page stand for two phases of a project. The former is a landscape planning with emphasis on ecology and water conservancy design. It includes repair of the best streamline of river bed, construction of multiple-water level adaptive river bed, as well as corresponding arrangement of multi-layer walking trail and gradient plant configuration matching water level. Besides, the planning employs multi-angle sectional design to respectively present upper river bed, lower river bed and corresponding plant configuration. In this manner, the sophisticated design intention and image are successfully delivered in the form of the simplest legend.

北京永定河门城滨水公园规划

本页图纸偏重于空间营造和景观城市设计，较为详细地表现了滨水公园区域的平面空间变化及功能分割，并通过片段式的图纸表达出区域内的典型空间，分别以鸟瞰形式表现场地的整体感觉及河床竖向分级，用人视点透视表达出不同层次滨水步道的实用感受。

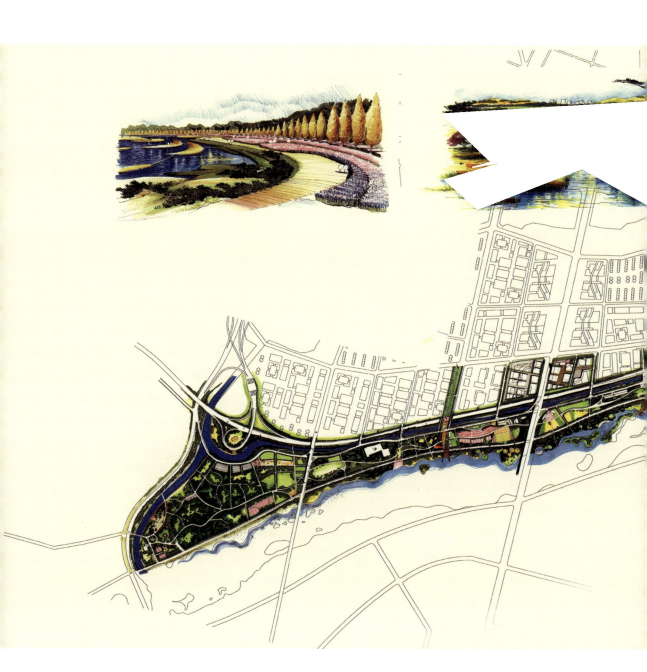

Planning of Mencheng Waterfront Park in Yongding River District, Beijing

This drawing lays stress on the spatial creation and landscape urban design, manifesting the plane spatial change and function segmentation of the waterfront park area in detail. The typical space inside the area is expressed via fragment-type drawings. The overall feeling and the vertical grading of riverbed is expressed in the form of aerial view, and the practicality of waterfront footpath of different layers is expressed from the perspective of people.

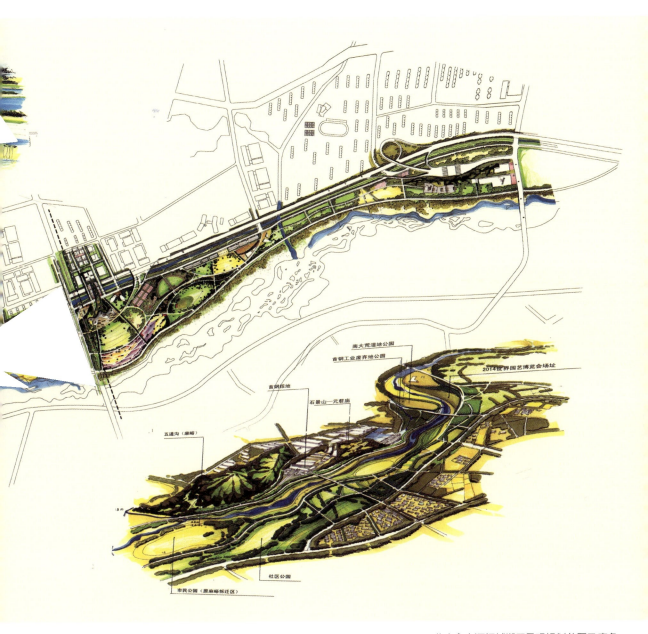

北京永定河门城湖区景观规划总图及意象
Landscape master plan and image of Mencheng

上海某住区规划

本页展示了一个典型的滨水住区的土地利用规划及相关景观意象表达,规划以水为魂,通过指状分布的土地划分以及尽端路为主导的交通组织,将三类密度不同的居住单元有机地关联成一个整体。利用指标从 0.5~1.5 不等的大中小居住单元,均具有面水景观和相对均好的景观,同时将主题组团用鸟瞰图的形式表达出整体的效果,同样以人视点的方式表达了使用者最切身的感受,达到了从整体到细节层面的兼顾和表达的清晰与具体。

Planning of a certain neighborhood in Shanghai

This drawing shows the land use planning and relevant landscape intention expression of a typical waterfront residential area. The planning takes water as the soul, and organically combines three types of residential units with different density into a whole via finger-like distribution of land subdivision and traffic organization led by cul-de-sacs. Meanwhile, the overall effect of the topic cluster is shown in the form of aerial view, and the most personal experience of the user is expressed from the perspective of people, achieving a combined consideration to entirety and detail as well as the clarity and specificity of expression.

农场酒庄景观

　　本页及后页所示 5 张小透视鸟瞰图,均是为前页的外部农场酒庄景观所做的意象图纸,毫无疑问原型来源于我们所熟悉的丘陵区域的欧洲梦幻农村,结合场所的特征进行了适当简化和元素重组,构成了一个崭新的清晰结构。其中的橄榄林、围合的大乔木、绿带以及点缀在田野花田之间的竖向的柏树,都形成了对场地构成的一种现实性的描述,这样的图纸同样应该围绕着总图周边,只不过因为上图图面空间已满,故而改用专题的形式陈列,这些快捷的表达途径、非常直接的图纸是对总体规划最直接有效的解释,其作用就如大规模科研项目中的附录,未必完全针对本场地,但对理解场地的真实环境氛围具有莫大的帮助。设计师应习惯并且多做此方面的练习。同样,这方面的图纸能够反映出一个成熟的景观师在常年的旅行、采风过程中所作出的积累性的感悟,很多主题性景观的设计过程就是在逐步表达的过程中逐步条理清晰和系统化。

Landscape of farm and chateau

    The five aerial view of small perspectives shown on this page and the opposite are the image drawings of the landscape of the external farm winery on the previous page. There is no doubt that their prototype comes from our familiar hilly region of the European dream countryside. It is simplified and elements are reorganized properly combined with the characteristics of this site, thus. It constitutes a new clear structure. The olive groves, the green belts, and the cypresses which are dotted amidst the field and the flower fields, develope a realistic description of site constitutes. Such drawings should also be around the general view, merely, because the space of the above drawing is full. So they are exhibited here with different topics. These fast and straightforward channels of expression provide the most direct and effective explanation of the overall plan; and its role is as the appendix in large-scale scientific research projects. It may not be entirely aim at this site, but it is quite helpful to the understanding of the real environment of the site. Designers should be used to doing more exercises in this aspect. Similarly, this aspect of the drawings can reflect a mature landscape architects' accumulation from the process of perennial travel and tour. The process of a great many thematic landscape design is the process of making the design clarified and systematic with the progress of design.

通过不同冷暖、明度的绿色系表达一个复杂的结构，将最深的色彩留给水，最鲜艳的色彩留给核心景区，对比最强烈的白色用来勾画道路结构，而周边最大面积的绿色山林则用最灰的绿色加以体现，形成由内到外，层次分明的结构序列。

A complicated structure is expressed via greens of different color tone and brightness. The darkest color is used for the water and the brightest color is used for the core scenic area. The color of white, with the strongest contrast, is used to draw the road structure, while the peripheral green mountain forest with the largest area is expressed with the greyest green, forming a well-bedded structure sequence from inside to outside.

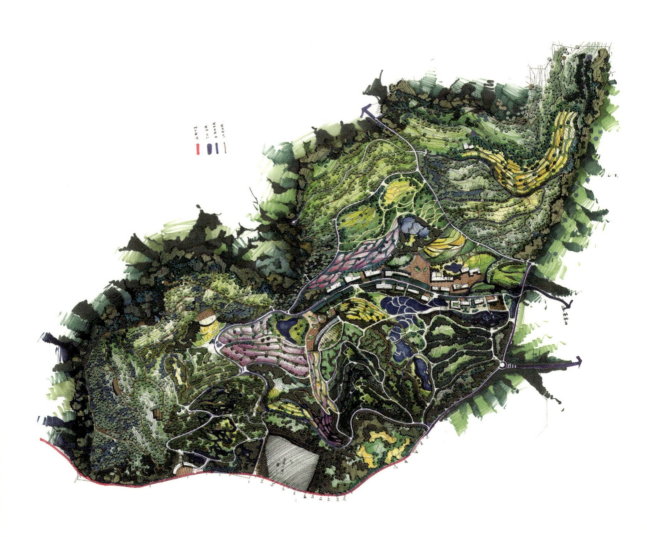

河北秦皇岛植物园景观规划

本规划处于两座山系的沟谷部分，地形设计极为复杂，需要兼顾植物园各个区域的均衡展示和现有山林植物品种的保护。在平面图设计中，使用了明显不同的色彩、笔触和植物单元去表达核心区域，缓冲区域及外围山林区域不同的景观特征，使规划的主体核心部分一目了然，同时用剖面图和透视结合的形式，表达了场地最核心的要素——大面积的沟谷和山坡以及上面丰富的植物配置。表达的效果充分，完备，设计表达关注了最核心要素的意向性表达这一环节。

Landscape planning of Qinhuangdao Botanic Garden in Hebei Province

The site of this planning is the gully region of two mountain systems. The landscape design is extremely complicated, requiring a comprehensive consideration to the balanced exhibition of each area in the botanic garden and the protection of the existing mountain plant varieties. In the process of plan design, distinctively different colors, brush strokes and plant units are used to express the different landscape features of the core area, the buffer area and the peripheral mountain area, making the core area of the planning as plain as daylight. The cores elements of the site—large areas of gully and mountain slope as well as the rich plant disposition on them, are expressed with the form of a combination of section and perspective drawings.

本页图纸均是为景观网络教学现场制作的综合表达图纸。平面图上通过色彩、形式的变化表达设计意图，言而未尽的部分均采用小透视的形式加以补充表达。这种小透视本身能独立成景，与平面图结合则具有更大的说明性，这是针对本章所说的综合性表达、一图多义的一个完整展示。

These drawings are all comprehensive expression drawings made during a landscape online teaching course. The design intention is expressed on the plan via the change of color and form. The unfinished part is expressed with the supplementary small perspective form. This kind of small perspective is able to form an independent scene by itself, which becomes more expressive when combined with the plan. This is a complete demonstration of the comprehensive expression and polysemy of drawing discussed in this chapter.

迁西滦河中央景观带入口广场设计草图

　　本页的图纸形式类似，均是采用不同尺度下，由远及近的表现场地的特征。从最简易的景观结构到全景鸟瞰，直至某个终点场所的细节展示，全方位反映场地的综合感受。这类表现对于草图空间的推敲，与业主沟通乃至设计团队的内部交流，都是简单易行而且清晰明了的方式。设计的延伸，修正和多种方案的选择都可以在作图过程中进一步加以完善。

Design sketch of entrance square of the central landscape belt of Luanhe River in Qianxi County

　　The two drawings on the left and right are of a similar type, both revealing the characteristics of the site from far to near under different scales. From the simplest landscape structure to the aerial view of the whole scene, to the demonstration of the details of a certain spot, the drawings reflect the comprehensive feeling of the site in all dimensions. This type of expression is a simple, practicable and distinct way for dealing with the spatial consideration of sketch, the communication with the client and the internal communication among the designer team. The extension and modification of the design as well as the selection of multiple schemes can be further perfected in the drawing process.

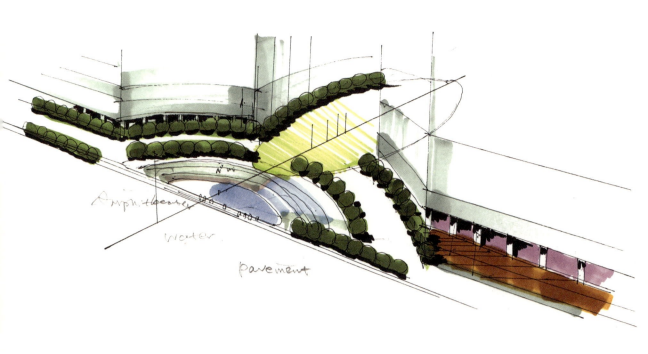

山西朔州七里河城市设计

选取场地核心中轴的不同视角表现同一主体。下图重在凸显全貌和城市的天际线特征，下图则主要表现中轴线内部特征，反映出一张由内向外的通景效果，两张图纸对照可以更为有效地表现出场地的设计特征。

Urban design of Qilihe in Shuozhou City, Shanxi Province

The same subject is expressed with different perspectives based on the core axis of the site. The below drawing pays attention to highlighting the overall perspective and the characteristics of urban skyline, while the right drawing mainly expresses the internal characteristics of the axis. These two drawings reflect a thorough landscape effect from inside to outside. The design features of the site can be revealed more effectively with the contrast between these two drawings.

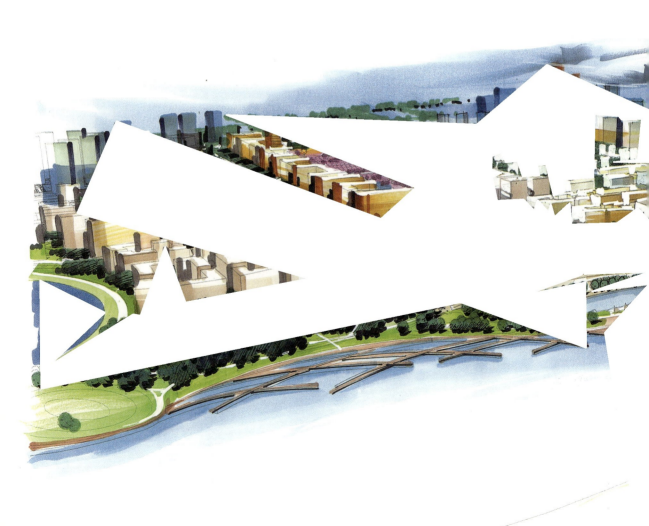

北京师范大学校园景观设计

在同一张图纸上并列场地最主要的三个区域，分别以透视、剖立面等形式展示外部形象，内部环境氛围和主要入口视角的景观，形成一种具有叙事性效果的连续画面，可以更为立体地表现出设计场地的环境特征、建筑形式、植物环境氛围、季相等多方面的景观特征。

Campus landscape design of Beijing Normal University

Three most important regions are shown in parallel on the same drawing. The external image, internal environment and the landscape of the main entrance perspective are shown via forms of perspective, section and elevations, forming a kind of continuous images with narrative effect, which makes it possible to reveal the landscape features in various aspects, including environment characteristics, building types, plant environment and seasonal aspect of the design site in a more stereoscopic way.

图书在版编目(CIP)数据

方案制图/王劲韬著. —北京：中国建筑工业出版社，2017.9
景观设计手绘教程
ISBN 978-7-112-21042-8

Ⅰ.①方… Ⅱ.①王… Ⅲ.①景观设计-绘画技法-教材 Ⅳ.①TU986.2

中国版本图书馆CIP数据核字（2017）第180815号

责任编辑：杜 洁 段 宁 李玲洁
责任校对：芦欣甜 王 瑞

景观设计手绘教程
方案制图
王劲韬 著

*

中国建筑工业出版社出版、发行（北京海淀三里河路9号）
各地新华书店、建筑书店经销
北京方舟正佳图文设计有限公司制版
深圳市泰和精品印刷厂印刷

*

开本：787×1092毫米 1/16 印张：9 字数：219千字
2017年9月第一版 2017年9月第一次印刷
定价：68.00元
ISBN 978-7-112-21042-8
(30672)

版权所有 翻印必究
如有印装质量问题，可寄本社退换
（邮政编码 100037）